Preserving Family Keepsakes

Do's and Don'ts

Preserving Family Keepsakes

Do's and Don'ts

By

Ilene (Chandler) Miller

Order from:
Ilene Chandler Miller
9712 Mirage Circle
Garden Grove, CA 92644-3023
(714) 636-3536

ISBN 0-938717-55-3

Second Printing 1996
Published by
Shumway Family History Services
5041 Stone Canyon Avenue
Yorba Linda, CA 92686
(714) 693-8703

This book is printed on acid-free paper.

TABLE OF CONTENTS

CHAPTER ONE

WHY PRESERVATION?

Family historians and genealogists understand the importance of family keepsakes. The preservation of these items is the responsibility of every family. We can fulfil this obligation by learning about the significance of preservation as it pertains to our heritage and those items that will help us perpetuate memories. Families are forever, but family keepsakes are not unless we take appropriate action.

Who would not cherish the thrill of discovering a very old trunk that once belonged to a great-grandparent, or that great-uncle who served in the Civil War? This has never happened to me, but I have often thought about the possibility.

The trunk, stored in a hot, humid, stuffy attic, has not been moved or even touched for at least one generation and perhaps two. It had been carefully stored in a corner, inaccessible to a casual observer.

I would be ecstatic to open that trunk and see items that were once treasured by a long departed loved-one. Just imagine, blowing away the dust and cobwebs to find a stack of old photos, which you did not even know existed, with names and dates neatly written on the backs. (Remember this is a fantasy!) What other items might be hiding inside?

It might contain a box of old love letters written by your ancestor to his young fiancee carefully tied with a piece of red ribbon. There is probably an assortment of newspaper clippings including obituaries of several dozen relatives. Your great-great-grandparents' ornate marriage certificate is there purposely bundled with their wedding portrait. The trunk might include a Confederate military uniform and medals. Perhaps a delicate linen tablecloth is tucked in there someplace. There might even be a handwritten family history including biographies of seven generations of your family, or there, securely wrapped in a baby quilt, is the family Bible with page after page of genealogy records.

The rush of excitement would be overwhelming. Once you compose yourself, you carefully reach in to touch the old, delicate, fragile items. One item at a time, you carefully remove them from their resting place. A tiny mouse peeks its head out of a pocket of the uniform and you squeal even louder than it does as it scampers away. Not to be discouraged, you continue. When you get to the items that have yellowed and turned brown with time and you hold them up to the light to read them. Slowly but thoroughly, they disintegrate in your hands. The remnants literally turn to dust as they are moved. The frustration is staggering. Your newly found treasures are useless!

After all that excitement, can you imagine the disappointment you might feel? In a desperate attempt to find the extent of the damage, you decide to continue going through the trunk to see what might be salvageable, just to discover it is nearly all a total loss. Only scraps, bits and pieces remain. There isn't even enough that can be identified to show to your family. In your anguish you might say, "I wish I had found the trunk sooner or not at all!"

How much of your family history and heritage would you be losing at the hands of time? What value could be placed on those treasures? There is no dollar amount that could be set to replace those keepsakes.

Saving something is not the same as preserving it. If all you want to do is to save it, you can put it in a box, store it in the attic or garage, and it will probably still be there in ten, maybe twenty, years. That is to say, unless fire, vermin, or a thief get to it first. But in what condition will it be? If a longer period of time passes, say fifty or more years, would your treasures be in any better condition than those found in the old trunk? Would your great-great-grandchildren have your treasures to enjoy and cherish in your memory?

What about that old shoe-box filled with letters dating from about 1850 that your family treasures? They detail the courtship of your great-great grandparents and his experiences in the Civil War. Maybe they have been read and reread over the years. With so much handling, the fragile pages have crumbled into a pile of postage-stamp size chips of paper. "Those old records had literally been loved to death."

When my mother was alive I spent many weeks, even months, arranging her photographs and paper keepsakes in a photo album. I neatly labeled each with interesting captions and arranged them in chronological order. It resulted in three very large albums that literally chronicled our large family in pictures. Later when she died, and it was time to divide her belongings, the other family members decided it

was time to divide her belongings, the other family members decided it would not be right for just one person to get all the albums. They divided them up by returning photos for each individual family back to them. I didn't agree but was in the minority. I got back all the photos I had given to my mother over the years (all duplicates of what I already had). The albums were disassembled and all that remained were miscellaneous snapshots. Those albums no longer have the same family significance and the chronology is gone.

But if we were to present awards for destroyers of the past, the winners would be people who intentionally throw out their records. At two critical times records run high risks of being thrown away: when someone moves and when someone dies.

Moving-time often triggers fatal decisions regarding records. At such moments old items, particularly if disorganized, seem like junk. The impulse to discard is especially strong for people moving into smaller quarters with limited storage space. Young people who leave home for college or careers often discard what they unwisely judge as 'silly' juvenile diaries and letters. Newly engaged couples gallantly (and foolishly) throw out bundles of old love letters.

Because of our mobility and the trend toward smaller families and smaller homes we have to be practical and selective about what we save. Do you ever think about items you or your family may have disposed of in your youth? If you had many comic books as a child you may have discarded them along with those stacks of old baseball cards, thinking they were just cluttering up your room. Who would have thought someday those items would be valuable?

"Death, particularly when unexpected, severely strains survivors who must make hurried and emotional decisions about the deceased's belongings. Widows and widowers throw out valuable items to remove precious memories that hurt too much. Children or grandchildren, when closing down a deceased parent's home, haul to the dump items not of interest to them, including letters and unlabeled photographs. If pictures are not labeled, or if they contain scenes and faces known only to the deceased, odds are high the photos will never survive."[1]

---

[1]William G. Hartley, *When in Doubt, Don't Throw It Out.* The Ensign, April 1986, p. 52-55.

Sometimes families just grow weary of saving old items. The widow of a California congressman tried unsuccessfully to interest local libraries in his personal papers. Their lack of interest convinced her the papers had no value, so she burned them.

WHAT IS PRESERVATION?

Webster defines preservation as *"To keep, guard, and observe. To keep alive, intact, or free from decay. To keep or save from decomposition. To prepare for future use. To keep up and reserve for personal or special use."*

Conservators, curators and preservationists further explain it as *"Whatever you do to an item, do not do anything that will permanently change it"*

" . . . must be reversible . . . "

"Never do anything that can't be undone . . . "

"It is better to do nothing, than to do the wrong thing."

"When in doubt . . . don't."

Because preservation involves chemical changes, it can be a very complicated subject. It can be extremely technical and involved. The purpose of this book is to simplify the subject so everyone can understand and take appropriate action. This is not a technical treatise—but where a more complex answer is warranted and where some people may be interested in a technical approach, that information is also included, in (small) footnotes, so only those really interested will even bother reading it. At the end of this book there is a glossary of terms and a listing of a few companies that sell archival supplies. It is important for you to know something about available products and what you need. We can easily be misled, by advertisements and well-intentioned, but ill-informed, clerks. A selected list of suppliers appears at the back. In using it, bear in mind that addresses can change in a relatively short period. Consult the yellow pages for additional sources in your area.

Any collection that has meaning and value for you, photographs, certificates, collectibles, fabric, newspaper clippings, and more—needs to be protected if you expect it to last very long. ***You can*** create a healthy environment for historic objects and maintain contemporary items in exquisite condition.

What is the difference between a collectible, a keepsake, an heirloom and an antique? A **collectible** is something that you collect either for personal enjoyment or as an investment. A **keepsake** is something kept or given to be kept as a memento. An **heirloom** is something of special value handed on from one generation to another. An **antique** is defined by Webster, "A work of art, piece of furniture, or decorative object made at an earlier period and, according to various customs laws, at least 100 years ago."

In this book these terms are used in a context meaning the same thing specifically as ***any object having family significance including those having special sentimental value to you.***

This book will take you through the Do's & Don'ts of preservation, and how best to display and store your valuables. Most of the information in Chapters One and Two apply to all keepsakes, although the first is focused on paper and the second is focused on documents. Many of the points covered in this book are things with which you have personal experience. We have all used cellophane tape to repair torn pieces of paper, old letters, or torn pages in a book. We have all used magnetic albums to store photos and scrap book items. We probably have photos stored in a shoe box or cardboard storage box in the back of a closet, in a dresser drawer, in the attic, cellar or storage shed. We have probably all written in ink on the backs of photographs and snapshots. As you read on in this book, you will learn these are some of the serious mistakes that are constantly being made.

Don't despair and don't be discouraged—we have all made mistakes—just learn the right way to treat your valuables so they will last. It is never too late to start preserving your collection. Preservation can be an expensive proposition but it doesn't have to be. There are ways to cut your costs and these tips in this book are marked with this symbol (cutting a dollar).

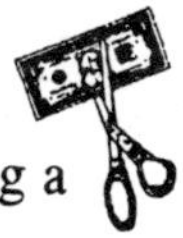

The information in this book has come as a result of much personal experience. Over the years, I have learned many hard lessons. I have spent lots of money just to discover I made bad choices in selecting methods of mounting and storing my personal family keepsakes. In many cases this resulted in major disasters and the loss of many items that I had wanted to preserve.

Working as a special librarian for over ten years I learned many techniques. Most of my expertise is the result of talking to experts in the field, reading everything I can get my hands on, and thoroughly researching the subject. I applied the information to the way I handled my own heirlooms. Even after all these years, I am still learning, because manufacturers are constantly improving and discovering new products that change old techniques.

Long before preservation was thought about seriously, I was lecturing on the Do's & Don'ts of Preservation. When I first started, archival supplies were almost unheard of by the average genealogist and family historian. They are much easier to find now. More and more companies are recognizing the public's demand for higher quality products but they still have a long way to go. Of course, these improvements bring higher costs.

I am primarily a professional genealogist and family history enthusiast. I have done this for most of my life. Genealogists, both amateur and professional, constantly search for better ways to gather information. We know the value of heirlooms and antiques that connect us physically with our forebears. Keepsakes allow us to reach out and touch our ancestors. The key to knowing our ancestors is to preserve those items we have in our possession that tell us about them. This can be in the form of documents, birth and death certificates, wills, military paper, tattered letters, *or* in items they once owned, a lock of hair, old photographs, personal diaries, old lace, linens or quilts.

Don't make the same mistakes I made in the past. Don't allow your family keepsakes to be slowly destroyed or lost entirely because of misinformation. Benefit from my experience and take the simple steps necessary to properly preserve your heritage.

CHAPTER TWO

PAPER

The one thing most of our genealogy keepsakes, precious documents and photos have in common is that they are all printed on some type of paper. Our generation takes it for granted. Our ancestors respected it. Paper is highly underrated. We need to develop a new appreciation for it. Paper has personality and character. We sometimes judge people by the quality of their stationery. Many of us like the feel and look of fine parchment and linen pages. I am fascinated with fine paper and can spend hours in a paper supply store looking at textures, colors and finishes.

How much do you know about paper? The quality of paper is determined by the kind and length of fibers it contains, the chemical composition of the pulp used, and the sizing employed by the mill.

All paper should be considered the same as a living thing. Paper breathes and moves. It expands and contracts according to the amount of heat and humidity it experiences. It does not like to be mistreated and has a long memory. Paper needs to be cared for. It needs tender-loving-care. We need to learn how to protect it from deterioration.

CAUSES OF DETERIORATION

What causes deterioration of paper? One obvious cause would be rough handling. The solution to that is obvious, we need to be careful how we handle a document we want to preserve! Keep in mind that paper is best when stored flat. It should not be rolled or folded.

When an item has been in existence for many years, it automatically turns yellow "with age." Or does it? The "yellowing" is caused by the presence of *acid*,[2] and has little to do with the passage of time.[3] Surprisingly, you will have greater success preserving older paper than modern paper.

HOW DOES ACID GET INTO PAPER?

No one in their right mind would intentionally put acid into paper, or would they? We all know that acid is caustic, so where does it come from? There are two ways: Internal factors (how it is made) and external factors (what happens to it later).

INTERNAL FACTORS

The internal factors are those things that go into the paper during the manufacturing process. The ingredients are what make such a huge difference in papers. This is what determines its type and durability.

HISTORY OF PAPER

When paper first came into existence, it was very rare and hard to produce. The cost was completely out of reach to most people. All paper was made by hand. Its chief ingredients were pulped linen or cotton and water. It was then sized[4] with a gelatine size for writing and printing purposes. Later, paper was commercially made from old rags, which were ground up and processed to produce a durable, high-quality product.

[2] Acid content is measured by its pH level. Webster defines pH as: "a measure of acidity and alkalinity of a solution that is a number on a scale on which a value of 7 represents neutrality and lower numbers indicate increasing acidity and higher numbers increasing alkalinity and on which each unit of change represents a tenfold change in acidity or alkalinity and that is the negative logarithm of the effective hydrogen-ion concentration or hydrogen-ion activity in gram equivalents per liter of the solution."

[3] Naturally there are other things that will damage paper, for example, household lye, and most caustic chemicals, etc. but these are rare compared to acid because they don't normally come in contact with paper.

[4] Size can be any glutinous material such as glue, flour, varnish, or resins, used for filling the pores in surfaces of paper.

As rags became scarce during the Civil War, there was a frantic search for a suitable substitute.

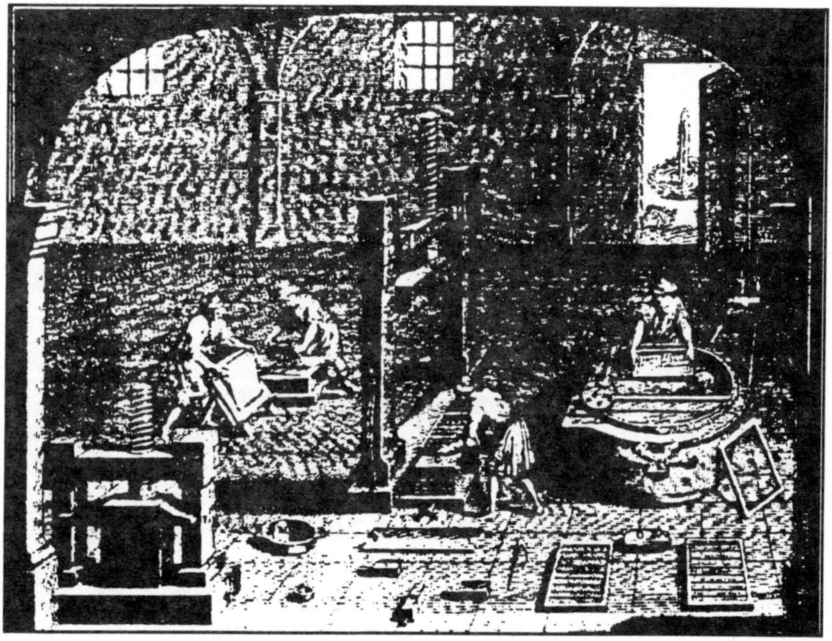

Early Paper Making

After the papermaking machine was introduced paper was made from brown wood pulp. Most of today's paper is made by machine. The pulp is ground or dissolved, usually by adding harsh chemicals (especially sulfur compounds) which leave an acidic residue. It is then rolled very thin to produce what we know as modern paper. The fibers in this paper are very short and the paper is not a good quality, but the cost is reasonable and paper is plentiful.

Rags are still used, but because of their short supply, cellulose fibers extracted from wood pulp is being used. Machine-made papers are also sized, although gelatine is no longer the primary sizing material.

material.

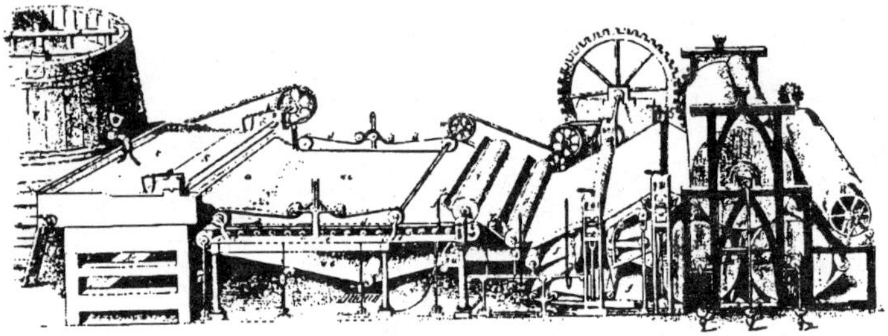

An Early Papermaking Machine

DURABILITY OF PAPER

Before 1800, the major cause of deterioration was not because of paper quality but inferior ink. Paper manufactured before about 1845 was high-quality, acid-free paper made from natural fibers. Since then, especially during the past approximately 100 years, the quality of paper has seriously diminished. Documents and paper items dating back 150 years are in good condition today. Items printed on paper produced since then have little chance of surviving on their own because of the high-acid content.

TYPES OF PAPER

An understanding of some grades of paper is helpful in understanding the quality of paper as it relates to the genealogist or home-archivist. Basically the wood pulp used to make paper is made up of two key ingredients, lignin[5] and cellulose.[6] Lignin is extremely

[5] Lignin is defined by Webster as *"an amorphous polymeric substance related to cellulose that together with cellulose forms the woody cell walls of plants and the cementing material between them."* Lignin is basically the key ingredient in wood products. It is extremely acidic.

acidic, cellulose is acid-free. Modern paper has a high sulfur content, because of the sulfuric acid added during processing. Alum[7] sizing is added and chemicals such as bleach and dye, among others, are also highly acidic.

The crudest forms of paper are used in brown paper bag and ordinary corrugated cardboard. When just a few other chemicals, especially bleach, are added during manufacturing it becomes lighter in color, resulting in paper like that used in inexpensive lined tablets used by children at school, or used in coloring books, in many paperback books and newsprint. This paper is highly absorbent and economical to produce. It is ideal for these uses but contains too much acid to preserve unless acid is removed.

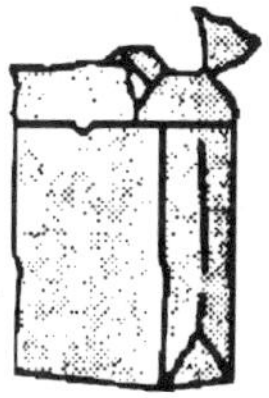

CARDBOARD

Cardboard is made from brown-wood pulp like paper but is thicker. Ordinary cardboard is made from the crudest form of paper and contains high levels of lignin and acid. Storage boxes are usually made from corrugated cardboard. Paperboard is a more refined form of cardboard. It is often bleached white or dyed. It is used in most boxes, including shoe boxes. All ordinary cardboard boxes contain massive amounts of acid. Cardboard contains more acid than any other form of paper. It is very dangerous to use if you want to preserve paper records for more than a few months.

[6]Cellulose is: "*a polysaccharide of glucose units that constitutes the chief part of the cell walls of plants, occurs naturally in such fibrous products as cotton and kapok, and is the raw material of many manufactured goods such as paper, rayon, and cellophane*" Cellulose is acid free.

[7]Potassium aluminum sulfate.

CELLOPHANE AND ACETATE

Easter baskets and fruit baskets are often wrapped in colorful cellophane. Frequently boxes use clear cellophane in the lid so you can see the contents. Toys are frequently boxed this way.

Acetate is what older page protectors were made from. It is very similar to cellophane but is more rigid. It is also used in packaging material. It is made from living cotton fiber pulp and acetone. This material breathess and thus has the ability to allow trapped odors and chemicals to escape or evaporate. Due to the porous or breathability of the acetate it is also brittle and tends to crack easily.

Cellophane and acetate are made from cellulose that is acid free, but chemicals and sometimes dyes were added in its manufacture that contain acid. This means anything made from cellophane can be acidic. Anything made from acetate will probably crack and will not last. You have probably noticed how these packaging materials often turn yellow and become very brittle with age.

Years ago, film holders and page protectors were made out of cellophane or acetate. If you have anything made out of these materials in your collection, you need to replace them before they cause any damage. Newer products made from acetate may be safe but need to be checked carefully..

TEXT PAPER

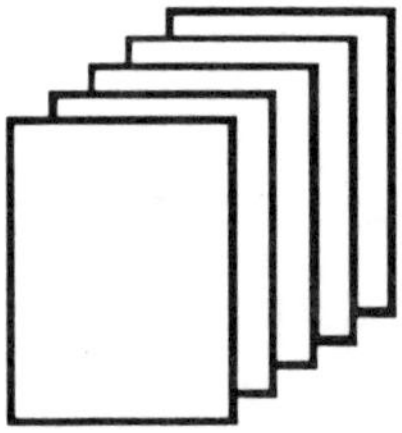

The type of paper we see most frequently is called text paper. During the manufacturing process even more chemicals, especially more bleach and sizing, were added to both whiten the pulp and give it a harder, less absorbent surface. This is the paper we commonly use in photocopy machines, on business forms, in most stationery, in ordinary books and in most things we think of as made of paper and cardboard. Colored text paper contains more acid than plain white paper. The dyes used are unstable and will fade quickly. In some cases colored paper will fade in a matter of days if stored on a shelf, exposed to room light.

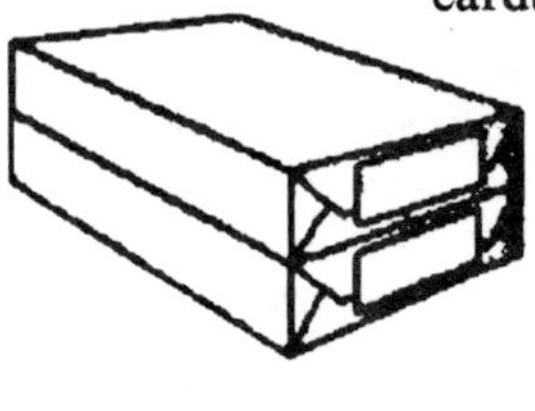

PHOTO-SENSITIVE PAPER

Paper that genealogists have used a lot includes paper that has even more chemicals added to the surface to make them photo-sensitive. These light sensitive papers are used in many older or less expensive copy machines. They are also the most common paper used in microfilm copy machines. These papers are very familiar and of particular concern to genealogists.

Only twenty-five years ago, this was the most common type of paper used in copy machines and thermofax copiers. Knowing this paper was heat and light-sensitive, I carefully stored important copies in a metal box to keep them out of the light and stored it in a cool place. One document, placed there for safekeeping, was a copy of a marriage certificate for a direct ancestor. I got the copy at a very small court house on a visit to a rural area of Southern Illinois. No one in the family had ever seen it before. I was thrilled to have it and took special precautions to preserve it.

Several years later, when I went to make a copy of the original, I was shocked! Much to my amazement it had dark brown, nearly black spots all over the back. Cringing, I thought, "Well, this page is the back side of the certificate, that is not too bad!" When I turned to the next page of the document, which was the front, my mouth fell open. It was totally black from the center out, obscuring all traces of the intricate designs that had once been there.

How could this have happened? It was stored properly. What I did not know at the time was how much acid there was in the paper. The document had self-destructed.

I wrote to the court house to get another copy. Several weeks later they answered saying the original record was no longer available. The document was lost to my family forever.

QUALITY PAPERS

It should be obvious that ordinary modern paper is not very high in quality. However, there are still many papers available that are made from natural fibers. Most of these papers will have a watermark on them showing the fiber content and are at least 25% rag or more. Other papers have been manufactured to meet the demand for better paper. If families across the country keep demanding better paper more will become available at lower prices.

ACID-FREE PAPER

Acid-free paper is any paper that contains an alkaline[8] base and is often lignin-free. This includes most natural-fiber papers that contain linen, cotton or rag fibers. Many bond papers and some copy machine paper is also acid free. It is durable and will not become brittle with age.[9] Nowadays this type of paper is commonly available.

Ten years ago acid free paper was quite expensive. Today, it is only a little more expensive than regular paper. The price difference is negligible. Not all paper mills produce acid-free paper. If the public were to express the desire for it, all mills would include it in their production line. Acid-free paper is also available in perforated computer paper.

Unbuffered[10] alpha cellulose paper is also available. This paper is ivory in color and is a beautiful acid-free product.

Check all the paper you purchase. If you are going to use it for genealogy or for something you want to last, buy the best paper you can afford. Never print a family history book on anything less then acid-free paper. Using acid-free paper, photocopy all originals that are too fragile to handle. It is ideal for creating permanent records, interleaving documents, wrapping, and lining boxes.

Acid-free paper alone is not enough to guarantee a long life to our

[8] Alkaline means it has a pH higher than 7 or there is an absence of acid.

[9] The pH of acid-free paper is neutral pH 7.0-7.5.

[10] Unbuffered acid-free paper has no chemicals added to protect it from future acid contamination.

books and documents. Although being acid-free will add many years to the life of your paper, it is totally dependant on how it is stored and with what it comes in contact.

ARCHIVAL PAPER

The first Archival Paper Ordinance was passed into law in 1907. It left a lot to be desired and in the 1960's it was revised and expanded. It stated all archival paper must be made from 100% textile fiber (cotton, flax or hemp). Even these papers, if improperly stored can become permeated with acid.

Archival paper is a step above regular acid-free paper. Besides containing no acid, it contains a buffering compound to act as a barrier to any future contamination from acid.[4] It is always lignin-free, and this paper possesses about twelve other qualities which not only make the paper feel and look nice, but assures a much longer life. Archival processing allows the paper to last over two-hundred years.

This paper is now available in a wide variety of colors. Although it is harder to find a source for these colored papers, they are wonderful to use as backing sheets in albums.

Archival paper is more expensive than other papers. Until more companies start producing it, the price will remain high, but it is far superior to all other papers and is worth choosing whenever you are the person selecting the paper.[5] Archival paper is NOT acceptable for color prints and fine silk and dye transfer. The buffering compound will damage color prints. It may be purchased from archival supply catalogs and some publishers. Most printing companies can order it direct from the mill.

---

[4]Buffered to pH 8.0-8.5 with 2% calcium carbonate.

[5]One company, Mohawk Mills, makes a fine text paper, called *Mohawk Superfine*. Artificial aging has indicated that it will not decompose in four hundred years. For information on how to obtain this paper, call Gary Shumway at (714) 693-8703. This paper is considered "top of the line." If you want the very best quality, this is the paper for you.

EXTERNAL FACTORS

What happens to the paper after it has been made are the external factors that happen and determine its acid content. Of particular concern is the environment—heat, light, humidity, dust, pollutants, and acid rain. They affect most fabrics, all paper products, including photographic paper and emulsions.

Extremes of heat and cold are very dangerous. As a quick guide, use your own comfort level to let you know. Harsh light with ultraviolet (UV) radiation causes fading. Swings in humidity cause substantial damage. The ideal humidity is 45-50 percent.

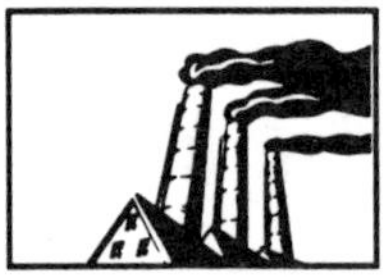

Air pollution is a real problem in our industrialized world. We hear about it all the time. We can see how it makes the sky gray instead of blue. We can feel how hard it is for some of us to breathe when air pollutants exist. But how does it affect paper and photographs?

WHAT IS ACID RAIN?

In the air we breathe there are chemicals that we cannot see or smell. These chemicals include, among others, sulfur dioxide and nitrogen oxide. What is not commonly known, is that when these two chemicals mix with moisture, they change into *Sulfuric Acid* and *Nitric Acid.* As this moisture settles to earth, it lands on literally everything, both inside and outside of our homes.

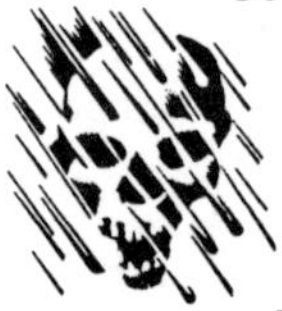

Look at the books on your shelves. Open one that has been there for a while, especially a paperback book or one printed on ordinary paper. You will notice yellowing around the edges of the pages. This yellowing is a result of acid rain. It is where pollutants in the air, dust and the acid in the atmosphere, have settled on the edges of the book. It has worked its way in toward the center of the pages.

WHAT IS ACID TRANSFER?

Acid transfer happens when acid migrates from one item or object to another. It is also called transmigration. Some books and items in our homes are made from good quality paper, with little or no acid in the pages. Acid transfer happens when something that contains a lot of acid comes in contact with otherwise clean acid-free pages.

A good example of this is when you place an object between the pages of a book. A rose that you want to press and preserve, or a book mark for example. Have you noticed that if you leave them in place for a while a stain develops on the pages of the book? This stain is usually acid-transfer and is yellowish or brown depending on the amount of damage that has occurred. Acid-transfer is also called *transmigration.*

THE THINGS WE DO TO PAPER!

We have all done things to our keepsakes that unbeknownst to us have caused serious problems. The old "paste & press" approach is dangerous. The "stick it in a box and put it in the closet" approach may eventually be regretted. We are our own worst enemy when it comes to paper.

You should always make photocopies of your original documents. Do this even if your original is a photocopy. Hopefully this copy will be made before any damage has occurred. Place your originals in a safe place and work with the photocopies to study, extract information, and so forth. Make copies on a copy machine that uses a carbon-based powdered toner and print them on acid-free paper. You will need to ask the person in charge of the copy machine about the toner.

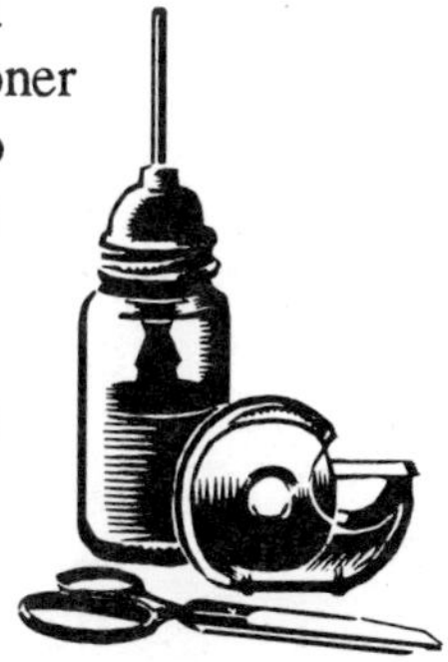

You should never attempt a "quick-fix," by using temporary measures. It is best not to grab the tape or glue to mend something. When caring for old documents, focus on preservation rather than repairs.

GLUE & ADHESIVES

Few things are as overlooked as much as the methods and materials we use to attach photos and miscellaneous items to an album page, mounting board, or other surface. These methods usually are done with some type of glue or adhesive.

As a rule, "any adhesive will eventually do some degree of harm." This is because nearly all household adhesives contain acid.

Manufacturers do not label glue and adhesives with a list of ingredients. Reading the label will only tell you what and how the manufacturer says it can be used. They must warn you about any harmful vapors, what to do in case it is ingested, and sometimes how it should be stored. They do not tell you how much acid the adhesive contains unless they want to sell it as acid-free.

So how can we tell how much acid is contained in the glue we buy? If the item is not marked, we must depend on word-of-mouth, archival supply catalogs and books like this. If you ask the seller they may or may not know (probably not).

Hundreds, possibly thousands, of adhesives are on the market, and for every type there are both good and bad products. Desirable characteristics are **stability, durability and reversibility.** Stability is important so that it will not break down and give unexpected results. Durability is its retention of adhesive qualities (it will continue to adhere). Reversibility is that the adhesive can be removed without damage to the material to which it has been applied and it can be removed with water. It would be labeled as, "removable with water," "washes out," or "water soluble."

TYPES OF GLUE

Webster defines glue as *"a hard protein chiefly gelatinous substance that absorbs water to form a viscous solution with strong adhesive properties and that is obtained by cooking down collagenous materials such as hide or bones."*

As a child I remember hearing about old horses being sent to the "glue factory." Glue made from animal hides usually needs to be heated to use and for that reason alone is seldom used by the public. Although not completely archival it is suitable for bookbinding. This type of glue is obtainable from some supply companies and hardware stores.

MUCILAGE:

Mucilage has been around for years. I remember using it as a child in school. It is made from organic material (usually seaweed) that is extremely unstable. As it ages, mucilage hardens and cracks and will eventually lose its adhesive ability.

- Don't use mucilage.

SPRAY ADHESIVE:

Most spray adhesives contain plenty of acid. They are highly toxic and extremely flammable. They work fast, they are strong and work well on cardboard, plastic, photos (according to the label) and more, but they will damage paper and fine photos in a short time. It is not recommended that they be used with keepsakes.

- Don't use spray adhesives.

RUBBER CEMENT:

Rubber cement is the most common glue found in stores. It was developed for use on paper because it will not wrinkle the paper. When it first became popular, I thought it was wonderful. It was so easy to work with and dried quickly. If excess rubber cement oozed from the edge of a photo or clipping being mounted, it could be easily rubbed off.

I liked it so much, not being content with buying the small bottles, I bought it by the gallon. After all, I had hundreds of photos and newspaper clippings to insert into family albums and scrapbooks! Using so much of it, a special thinner was required, because the rubber cement would become too thick to use. The special thinner had fumes that were overwhelming. I thought that was a small price to pay for the convenience of this great product.

Of all glue available, rubber cement is the most acidic, therefore the *most dangerous* adhesive you can possibly use. Not only the glue itself, but even the fumes are dangerous.

Like I said in the introduction of this book, this was one of the hardest lessons I learned over the years. I had no idea what the rubber cement was doing to my keepsakes. Use it *cautiously* on temporary paper projects only. *Never use it on anything you want to last more than a few weeks.* Damage occurs quickly and could affect not only the item it is used on but also adjoining pages.

- Don't use rubber cement on important items.

FLOUR & WATER PASTE:

Remember when we were children and our own glue from flour and water? This was actually a very safe, good, acid-free glue. Pure starch paste—wheat or rice—meet the requirements of good adhesive characteristics and have held up for over three centuries. The major problem with it is the flour attracts insects. It also needs to be refrigerated after mixing and may form mold if stored too long. It should not be used on anything you want to preserve.

- Don't use unless you can be sure no insects can get to it.

METHYL CELLULOSE:

Methyl Cellulose paste is ideal for conservation mounting. It is pH neutral, the powder simply dissolves in water, and it is vermin-proof. It can be redissolved in water if you need to remove it. Methyl Cellulose keeps indefinitely after mixing. If it gets too thick, just add water. Librarians frequently use this. It is the ideal paste, if you don't mind having to mix it with water to use it. However, it is not available to the general public. This paste is usually called Methyl Cellulose. You can obtain it only from archival supply or library supply catalogs, not department stores.

WHITE GLUE:

White glue[13] is called "America's Favorite Glue." It was one of the first synthetic glues on the market. It is for all porous material, it is nontoxic, dries clear, fast and strong. If you read the label very carefully, they add a note saying "Not for photos." What they do not tell you is that this glue contains plastisizers, which is highly acidic and unstable over a period of time. It dries hard, will crack and becomes brittle. This glue is washable until it dries, then it becomes rigid. It is good for woodworking but not for any paperwork that you want to preserve.

- Don't use regular white glue on paperwork.

POLY-VINYL ACETATE:

PVA (PolyVinyl Acetate) white glue is ideal for bookbinding, box making, collages, even hard-to-hold plastics. It dries clear and forms a strong, long-lasting bond. It can be several times more expensive then Elmer's-type white glue. PVA and Elmer's look identical, but PVA remains flexible and will not crack or become brittle. It is considered to be almost a completely safe glue.

There are two serious problems with this glue. If PVA ever freezes while in its liquid state, the glue is ruined. It becomes watery and loses its effectiveness. This glue is also not reversible.

This glue is the type used in making many crafts. You probably cannot purchase it at your local discount store. You might find it at craft stores (a common brand name is Sobo glue), also at library supply and archival supply stores (often called PVA glue or White Neutral Adhesive).

- Don't use PVA on most documents; it may cause wrinkling.
- Do use PVA for making boxes and other archival projects.

COMBINED GLUES

Combining PVA and wheat paste or PVA and Methyl Cellulose produces a less expensive mixture that works well on books and most keepsakes. Experiment with the formulation for best results.

---

[13] Elmer's glue, made by Borden, Inc., is an example.

GLUE STICK

Glue stick is not completely acid free but the water-soluble type is considered non-acidic.[14] It is the safest glue that you can buy at your local store. You still need to be careful where you apply it. Do not allow it to touch the face or front of a photograph since it can leave a stain that is nearly impossible for amateurs to remove. Apply it to the back edges of photos and paper items only. There are four types of glue stick.

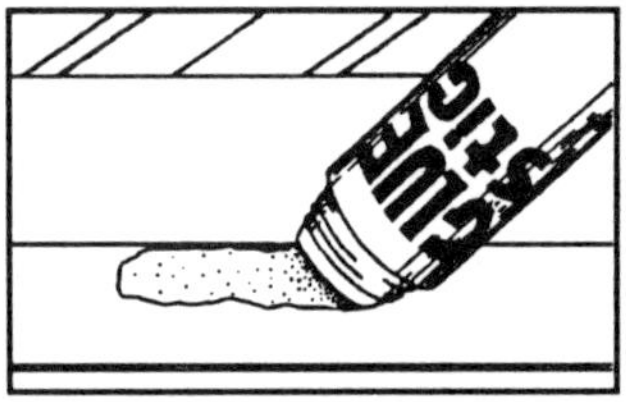

- Clear glue stick, the type that is ***not*** water soluble should be avoided. Read labels carefully. If it requires something special to clean it up or remove it, don't use it. Know brand names that are safe. Never use on items for permanent retention.

- White glue stick which is water soluble and is made from white paste. It is ideal to use for paper-to-paper applications and with lightweight fabrics. Brand names include Dennison, Scotch (3M),[15] Faber-Castell (now Eberhard, Faber Co.)[16] .

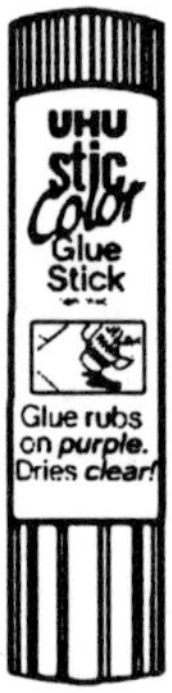

- Purple glue stick is the same as white glue stick but is purple when it is applied to show where you've placed the adhesive. It dries clear in a short time. This allows you to quickly clean up any excess glue before it dries. It is made by most of the companies named above.

---

[14] Glue stick has a pH of 12 $\pm$ .5.

[15] The 3M company makes no claim that this product is acid-free or safe for archival mounting.

[16] Eberhard, Faber Co. makers of UHU glue stick guarantee their products.

■ "Post-it" type Glue Stick[17] remains tacky so you can rearrange items before attaching them permanently. Never use it on the face of a photo. This is the same glue used on Post-it Note Pads. It is safe to use, if used with caution. All glue sticks can be purchased from department stores, discount stores, and stationery stores.

TAPE

Many tapes will turn yellow in a very short time. In most cases, the adhesives that hold the tape in place are corrosive, leave a residue, and are not reversible (one of the cardinal rules of archival preservation).

Never use tape over the front, edges, or corners of a photograph or document. Tape does not allow the paper to breathe. Once acid migrates from the tape to the paper, the tape will not adhere to the page. Have you noticed how once tape comes loose, it falls off the paper, leaving a yellow mark? This is because the acid in the adhesive has transferred to the page, causing the paper to yellow and tiny areas on the paper start to dissolve.

TYPES OF TAPE

CELLOPHANE TAPE:

Cellophane tape, as it ages, becomes brittle, turns yellow, and comes loose from the page after a short time. It is primarily the adhesive that makes this tape dangerous. In addition, the surface that you apply it to is permanently damaged because you cannot remove it. When opening gifts that were sealed with tape, I am sure you have noticed how the tape destroys the wrapping paper. It removes the design on the paper. When you apply it to a document you want to preserve, it would do similar damage to the printing or ink.

[17]Other brands call it "Tack-a-Note" by Dennison, "Post-It" by Scotch, etc. They have a pH of 10 ± .5.

- Don't use cellophane tape on important paperwork.

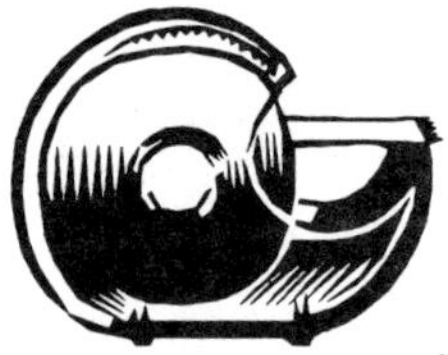

SCOTCH MAGIC TAPE:
Manufacturers of cellophane tape became aware of the yellowing problem and developed what became known as Scotch Magic[18] tape or non-yellowing transparent tape. It is now known as Magic tape. This is now the most common tape found in stores. It has a dull finish and you can write on it. These companies make a fortune on this product. However, don't be deceived! The problem with this product is not the tape, it is the adhesive that contains acid and causes discoloration.

- Don't use non-yellowing transparent tape on paperwork you want to save.

VINYL TAPES:
Vinyl tapes are very common. They are used for a variety of things, including electrical repairs. The most common color is black, but they are also available in many other colors. Vinyl is a petroleum-based product and contains an oily residue. Combine this residue with the acid from the adhesive, and it can cause lots of damage to paper and books.

- Don't repair paper or books with vinyl tape if you want them to last more than a couple of years.

MASKING TAPE or BROWN SHIPPING TAPE:
Masking tape is made using inexpensive paper with a highly acidic adhesive. This paper will disintegrate quickly. It was not designed to last very long. Have you ever used masking tape to paint something

[18] Scotch and Magic tape are trademarks of the 3M Company.

and forgotten to remove the tape immediately just to go back later and find the tape is very hard to remove? When you do, you find it removed the painted surface underneath!

Brown tape is made from kraft-paper. It is fine for shipping, but not for long term storage and should never be used to seal storage boxes.

- Don't use masking tape on anything to be used more than a few days.
- Don't use brown paper tape on anything you want to save.

CLOTH TAPE:

Cloth tape was designed to last longer than paper tapes. Because it is made from cloth, it can withstand lots of handling and wear. It stands up for quite a while, but again, it is the adhesive that causes damage to the item being taped.

- Do select a cloth tape that has an acid-free adhesive.

REMOVING UNSAFE TAPE

How do you remove unsafe tape? You can't, without damaging the surface. A professional conservator can do it but it requires special equipment, special chemicals and lots of experience. Keep this in mind when you use acidic tape.

Some tapes can be removed with a small microspatula, a special tool for lifting and scraping surface debris from documents. It should be made of stainless steel with one end tapered, one end rounded. This tool is also handy in removing photos from album pages. Removing the adhesive requires the expert. The most common chemicals used include benzine, acetone and toluene, all extremely toxic.

Once the tape is removed, any damage that previously occurred to the paper will remain. For instance, if the paper has already turned yellow from acid transfer, the paper will still be yellow when the tape is removed.

MENDING TORN PAGES

One tear can cost over one dollar to mend. Archival mending of pages is done with acid-free tissue impregnated with wheat paste and treated to resist insects. It is clear enough not to obscure type on the page. To apply this tissue, it is not cut, but is wet-torn. This separates the fibers instead of cutting them, which leaves a smoother, flatter edge. It is barely noticeable to the person reading a book.[19] A very thin special-purpose acid-free tape is also available in a one-inch width, which is less expensive than the above tissue. It does a good job repairing torn pages in books and on documents. You can purchase it at library supply and archival supply stores and through catalogs.

SAFE TAPES

A safe tape would be any tape made from Mylar"D" or Scotchpar film or other acid-free material coated with a pH neutral adhesive. These tapes are available from some art or library supply stores or through archival supply catalogs. Most come in a variety of brand names. These include:

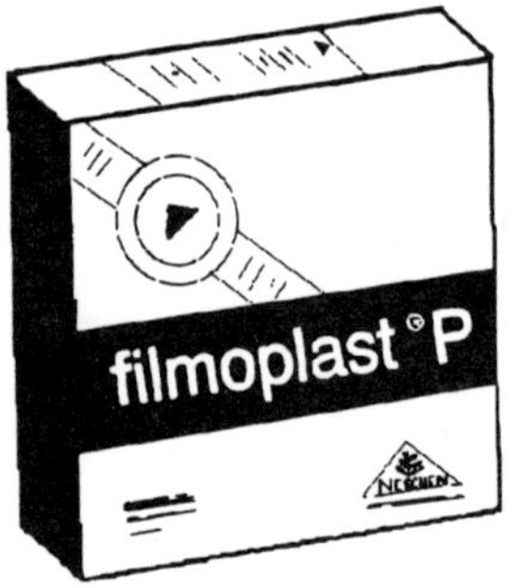

- Filmoplast[20] —a nearly transparent pressure-sensitive tape, available in several different weights. Distributed by the Filmolux Company.

---

[19] One sheet of this treated mending tissue (12x18") can sell for about $8. It includes blotters, nylon and instructions.

[20] Filmoplast and Filmolux are registered trademarks of Neschen International, made in Germany. These tapes sell for about $20 (or more) for each 7 oz. roll. That is enough to do a lot of mending.

- Non-woven polyethylene—pressure-sensitive adhesive.
- Linen tape—acid-free, water activated adhesive.
- Paper tape—acid-free & buffered, water activated adhesive.
- Double sided (coated) transparent tape #415 used in encapsulation (made from ScotchPar polyester by 3M, uses an acrylic adhesive). Other brands of double-sided tapes may not be safe.
- Polyester transparent tape #850 (same as #415 but has adhesive on one side only).

LABELS

Ordinary labels use adhesives that are extremely dangerous for paper and documents. The adhesive is similar to that used on most household tapes. After a few years, the adhesives becomes brittle and transfers completely to the paper and the label falls off. This includes most Avery and other brands of self-sticking labels and older lick and stick labels..

Acid-free labels are available that are made from acid-free paper and contain a permanent acrylic adhesive. They are acid-neutral. These labels stick to almost any paper, plastic, glass, Mylar D polyester, polypropylene, backs of photographs or metal surface. They withstand projector heat. They should never be used directly on the face of a document or photograph.

Also available are foil-backed labels. They are coated with a special formulation of the acrylic adhesive. The unique foil barrier prevents surface inks from penetrating the label and affecting the material beneath the label. They are good for use on boxes, folders, spines of books and back of photographs.

Labels are available in pin-fed single rolls and sheets for use in typewriters and dot-matrix printers. They are also available on flat sheets with four or five across depending on the width. Labels should have rounded corners for ease in removing. Acid-free labels are available from archival supply catalogs.

FADING AND DISCOLORATION

Every time we look at our keepsakes we are exposing them to light. Paper, textiles, color photographs and ink will eventually fade and deteriorate unless they are protected from damaging light. Sunlight can be devastating to keepsakes, not only because of ultraviolet rays (UV), but also because of visible light. That means all light, not just sunlight, will cause damage. Light from both tungsten and fluorescent light bulbs can cause deterioration. The least to the most damaging light (comparable brightness) is:

- incandescent (tungsten) light (no UV)
- halogen light (no UV)
- fluorescent light (UV)
- indirect daylight (UV)
- direct sunlight (UV)

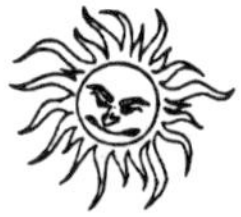

Museums use special filters and lights, but they are very expensive and not very practical for home use. Plexiglass[21] prevents UV rays from penetrating paper items you display. Some plastic sheet material emits harmful substances and may damage photographs. The ideal solution would be UV treated glass.

Discoloration usually occurs when chemicals or dyes in the paper, cloth or photo start to deteriorate. It can also happen when a colored object is placed next to the item and the chemicals and acid from the foreign object transfer into the item you want to preserve. If the discoloration is a color other than yellowish-brown, it is probably from the dye or ink in the foreign object which has transferred.

Once something has faded or discolored, it cannot be restored to the original. The damage is irreversible. All keepsakes should be protected from light. Using photographic steps such as polarized and infrared techniques, you can recover some faded images or writing by making a copy negative.

- Don't expose items to any more light than necessary, the darker the better.

[21] Plexiglas is a brand name for a clear acrylic sheet of plastic.

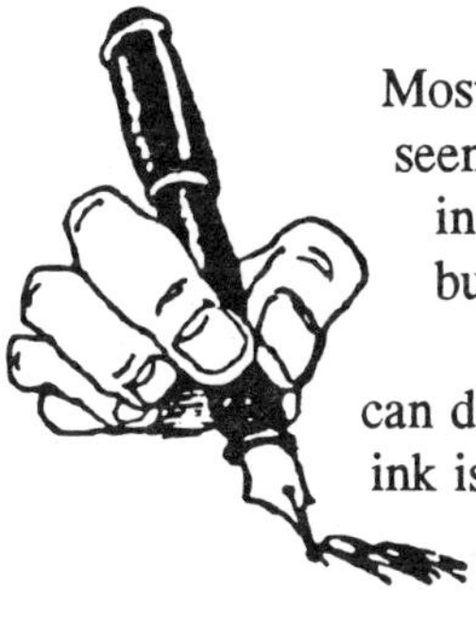

INKS

Most ink fades, bleeds or transfers. We have all seen how ink, especially blue or colored ink, fades in a short time. Black ink is usually more stable, but will still fade depending on how it was made.

Common inks contain a lot of acid. Most ink can damage the paper it is on. The formula[22] for the ink is usually not water-soluble and is what makes the ink stable and/or permanent. Even, so-called, water-markers contain solutions to make them stable and make the ink difficult to remove.

We all know how hard it is to remove ink stains from clothing and furniture. Likewise, ink cannot be removed from the backs of photographs (remember the definition of preservation...it must be reversible!). Because of the acid in the ink, if you view it through a microscope you can see the damage it does to the paper.

Depending on the chemicals used in the ink, it can bleed through the paper onto the opposite side. This is especially true of markers. Permanent markers are the worst. Try writing on a piece of ordinary copy paper with a wide permanent marker. It instantly soaks through the paper. This also happens to photographs and can ruin a beautiful picture instantly. Ordinary ballpoint pens fall into this same category.

Ink can also transfer. This means if you write on a piece of paper, even if you allow the ink to thoroughly dry or set, it can transfer to an adjoining piece of paper, book cover, or page protector (unless it is the non-transferrable type).

India ink is a fairly stable, good ink. You may have used it on photos or in albums that you have owned for many years. It probably remains in good condition. Ink used by calligraphers is also a good, stable ink. They are not acid-free however and are not reversible.

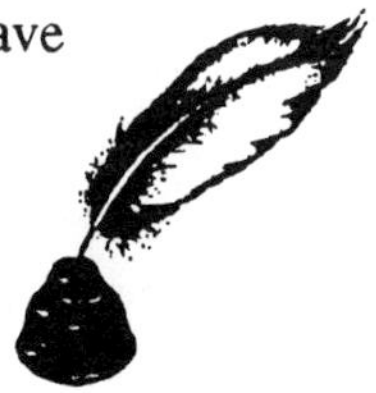

[22]Includes a non-aqueous medium.

To be archival, inks must be resistant to light, heat, and water. They must not contain any impurities that would affect the permanence of paper, photographic images, or storage containers, and must not deteriorate with time.

Acid-free ink is not easy to find. At one time the only place it was available was from the Preservation Office at the Library of Congress. Since then a few inks have been developed that are safe to use. Some are not recommended for technical drafting pen use and usually must be used with a dip pen or on a stamp pad. However a few companies offer this type ink in ballpoint pens selling for about $5 each. All ball point pens should be used cautiously. The pressure of the pen tip can cause an indentation on some papers and especially on photographic paper. Felt tip markers and micro-point pens are better but must contain a pigment based ink.

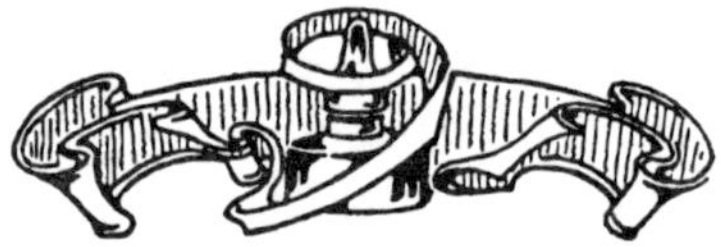

Archival ink is permanent and is often called actinic ink or pigmented ink. Once these inks are dry, they are waterproof. Caution must be used to guard against contamination of pens and stamp pads by other non-archival inks. Actinic and pigment inks are available from companies that sell preservation supplies.

Printing inks require time to set. No deacidification, lamination, or encapsulation should be attempted until at least a year after publication. When making photocopies of anything, use a machine that uses powdered toner (if in doubt ask someone about the toner). This toner contains carbon and is permanent. It remains to be seen how long this will last however as there may be a problem of flaking if the toner is not properly fused to the paper.

- Do use a stable acid-free ink on documents and paperwork.

CHAPTER THREE

PRESERVING DOCUMENTS

How important are your documents? Are they originals? Can they be replaced? Do they contain original signatures? Are they worth the cost of preservation?

Decisions to preserve documents in their original form should be made carefully. The cost of preservation treatment and continued housing of the materials should be weighed against such factors as the subject content and the availability of the information it contains. If you decide to keep it, make a photocopy of it. Copies are considerably less expensive and if they are printed on acid-free paper, they will last much longer. In the case of a newspaper clipping you might decide to discard the originals.

Now that you know about paper, you have a good idea what is going to be needed to preserve your documents. First you need to find out if there is any acid in it. If there is none, you can go ahead and display it or store it taking precautions not to contaminate the document by improper handling or storage.

If you discover acid is present in your keepsake or document, you will need to neutralize the acid before you display or store it.

PRESERVING NEWSPAPER CLIPPINGS

Newsprint characteristically contains short wood fibers and impurities. It is manufactured from acidic pulp, and has a low surface finish. Initially, fifteen to twenty percent rag pulp was added to increase the paper's strength. As the demand for paper increased and rags became scarcer, the rag pulp was gradually replaced by sulfite pulp. The result was an inherently weak and unstable paper. Yet newsprint continues to be popular because of its low price and high absorbency, which makes it an economical and excellent medium for high-speed presses. Because of its instability, however, it cannot be preserved for an indefinite period without special treatment (as stated on page 33).

Most clippings, whether individually housed or mounted in scrapbooks, have already begun to deteriorate and become discolored and brittle, sometimes so brittle that pieces have broken off or crumbled.

Unmounted clippings should be deacidified and given an alkaline reserve to arrest further deterioration. They may then either be encased in polyester film (see page 47) or placed in a scrapbook with alkaline covers. If clippings have been mounted previously, they may need to be removed from the backing, then treated.

Printing inks require time to set and no treatment should be attempted until at least a year after publication. During this year, the clippings should be stored in acid-free folders, in the dark, and at the lowest feasible temperature.

Once the acid is removed from clippings they may be left in the original scrapbook if the backing is sound. For added protection, each leaf may also be encased in polyester. In either case, clippings should be mounted on acid-free paper, using a neutral adhesive.

Occasionally you may decide you want to preserve an entire newspaper or pictorial. Follow the same steps for removing acid as for clippings. Then store the treated newspaper in an archival storage box large enough to lay it flat.

TESTING FOR ACID

How can we identify the presence of acid in our keepsakes? In much the same way you would test a swimming pool for the presence of acid. You use a pH testing kit or a testing pen, specially manufactured to test for the presence of acid in paper, envelopes and mat boards. The pen is an easy way to test the acid content of materials. The cost of the pen is only a few dollars (ranging from $2 to $4 each in most archival supply catalogs).

The pen looks like an ordinary felt marker, except the tip is not colored.[23] You touch the tip to the paper in an obscure spot so it cannot be easily seen (like to the back or to an edge or corner) and make a tiny dot. A color, ranging in degrees from yellow to green to blue, will appear. If it turns yellow it means the paper has a high acid content; if it turns green there is some acid present; if it turns blue it is acid-free. The dot remains on the surface to serve as an indicator of future contamination. The dot changes color if the pH of the paper changes.

DEACIDIFICATION

Once it has been found that your document contains acid, it should be treated as soon as possible to prevent damage from occurring. The process of neutralizing the acid is called ***deacidification***. This term is also taken to mean not only neutralization but also alkalization, which means a buffering compound is deposited.

Before encapsulation or framing, documents, newspaper clippings, and works on non-archival paper should be deacidified to keep them from yellowing and deteriorating. Impregnating a high-alkaline reserve into the paper neutralizes existing acids and prevents the development of future acids.

Printing inks require time to set and deacidification should not be attempted until at least a year after publication. Treating documents with these chemicals can be expensive and time-consuming. It is important to consider the cost and your time, versus the value of the items you want to preserve.

If you want something preserved, but you do not want to do it yourself, consider having it professionally done. I have been doing this type of work for fourteen years and would be happy to consult with you. *(Keepsake Photos, 9712 Mirage Circle, Garden Grove, CA 92644, phone 714-636-3536)*

[23]The pen contains Bromcresol Green, a chemical ink that changes color.

Deacidification **does not** restore deteriorated paper to good health, but it does stop further deterioration in paper that has begun showing detrimental signs.

DO-IT-YOURSELF DEACIDIFICATION

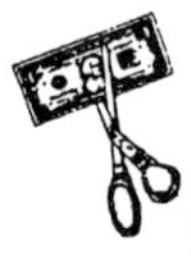

Once your decision to preserve clippings or documents in their original form has been made, the least expensive method is to learn to do it yourself. Before you attempt to do this, be sure to make good photocopies on acid free paper or archival bond paper.

TWO METHODS

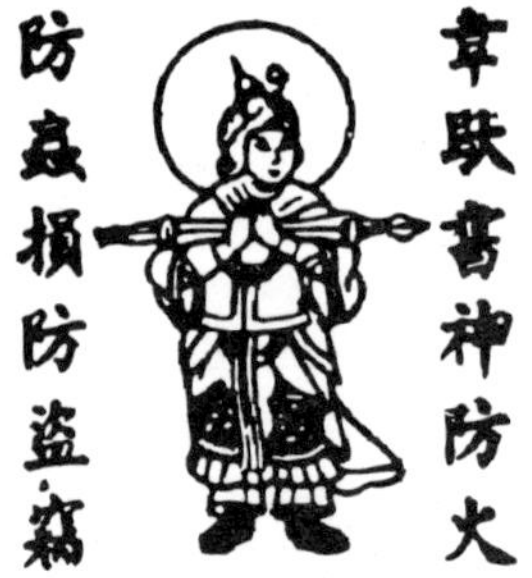

Wei T'o
The Chinese
God of Paper

There are two basic methods of removing acid from paper.[24] The method preferred by professional librarians and archivists uses a non-aqueous solution. It is an expensive process. At one time, it could only be done by professionals, because of the availability of the solutions.

Pure preservationists do not approve of the second method, using a simple home-recipe, but they admit it does work. There are several disadvantages including: (1) The water in this recipe can damage some paper[25] ; and (2) It is difficult to work with and takes longer to dry. The advantage of method #2 is the nominal cost.

---

[24]Results may vary for both methods. There is no guarantee that these procedures will not harm your keepsakes. There are many variables to consider.

[25]If the paper is colored or has color in it, this solution could make the color run depending on the printing ink or dye used.

METHOD #1

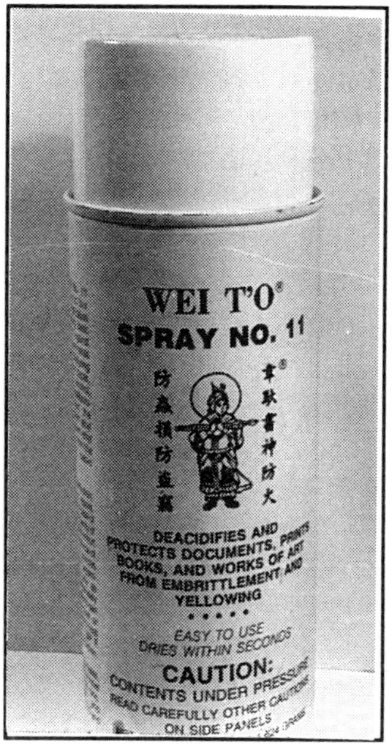

This method uses solutions that contain an alcohol base. There are two available: Wei T'o[26] and Bookkeeper.[27]

Wei T'o is the name of the Chinese God of paper. It is also the name of a spray that deacidifies and protects documents, prints, books and works of art from embrittlement and yellowing. Wei T'o is available in liquid form or in a spray container.

This solution extends the life of acid paper two to four times by (1) neutralizing the acids that cause paper embrittlement, (2) depositing basic magnesium carbonate to protect against reacidification.[28]

Another solution, which is considerably more expensive, is available. It also contains no CFC's (ChloroFluoroCarbons), and comes in a spray pump bottle or aerosol can. One brand name for this is "Bookkeeper." (It is considerably more expensive than Wei T'o.

Spray solutions are recommended particularly for persons with limited equipment due to its low vapor hazard. Liquid solutions are also available but require careful handling (similar to handling techniques explained in Method #2).

---

[26]Wei T'o is a registered trademark of Wei T'o Associates, Inc.

[27]Bookkeeper is the registered trademark of Preservation Technologies, Inc.

[28]These solutions also inhibit trace metals (iron, copper, and cobalt) from catalyzing oxidative reactions which discolor paper. The solvents, denatured ethyl alcohol, chlorodifluoromethane, and trichlorotrifluoroethane, dissolve and transport the deacidification agent, a magnesium ethyl carbonate, into paper. This magnesium ethyl carbonate reacts naturally with air to form the benign, long term protector, basic magnesium carbonate. Since Wei T'o no longer contains CFC in the formulation, the cost has risen.

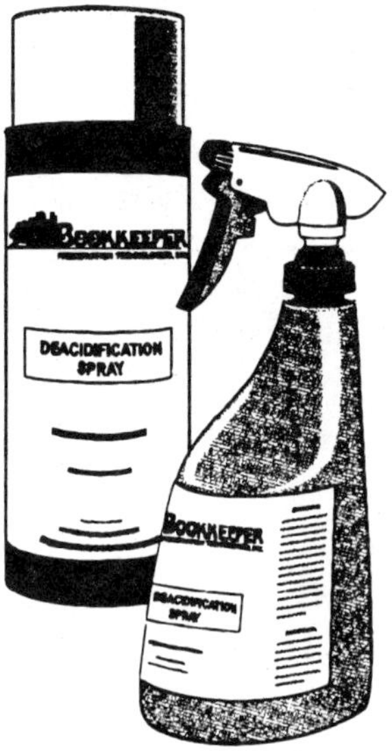

MATERIALS NEEDED:

- **Wei T'o or Bookkeeper spray.**
- **Blotting paper**
- **Erasers[29] & dry cleaning pad[30]**
- **Cotton swab (Q-tip)**
- **Spray-board rack[31]**
- **Sheets of waxed paper or silicone release paper[32]**

---

[29]A kneadable eraser works well, or a soft eraser.

[30]These can be purchased at many art supply stores or where they sell archival supplies. Use the type designed for documents *not* the type used by draftsmen (it is too harsh). These dry cleaning pads are available in pad form (recommended) or as a powder that is sprinkled on and dusted off the paper. The pad is the most convenient.

[31]A spray-board rack is a non-porous board or rack that will hold your document or paper to be treated in an upright position (tilted 110+ degrees or the angle of an easel). This is for ease in spraying. Be careful the document does not bend or sag. This is optional, you could just lay the document flat.

[32]A special paper that is treated with silicone so it will not stick. It is also useful when recently pasted work requires pressing while still damp.

DIRECTIONS FOR SPRAY TREATMENT:

① Inspect document and prepare it for treatment.

② Test it for sensitive inks, make sure they will not run or smear. If you should notice running, blot it carefully and quickly. Wei T'o and Bookkeeper are safe for most inks, but all inks should be tested. Use a cotton swab soaked in Wei T'o solution and gently roll the swab over the ink in an obscure area.

③ Remove dust, and any other surface dirt or marks. Use a soft, clean eraser or a kneadable pure rubber "groomer." This is permanently tacky, pH neutral and long-lasting. It removes pencil marks, charcoal, grease, debris and much more. Always use erasers in one direction and never over writing on a document.

Also recommended is the use of an opaline[33] dry cleaning-pad or powder (<u>not</u> the type draftsmen use, it is too harsh). This will help remove the surface dirt, even dirt that is not noticeable to the eye, like smoke film. Work cautiously on pencil. It is better not to use it on chalk, charcoal or pencil drawings. Do not rub or roll, simply dab the surface.

④ For best results, place the document in an upright position on a spray-board-rack (or piece of heavy cardboard covered with waxed paper). Spray according to the directions on the can. Wet the document thoroughly and evenly by spraying from side to side in an overlapping pattern. Do not let the liquid drip or run. Start on the back with a heavy coat, go to the front with a light coat.

⑤ Lay the document flat to dry. If the document cockles (wrinkles or puckers), complete drying between sheets of dry, white blotting paper.[34] Apply moderate but even pressure (a piece of glass works well). It usually dries quickly (about one minute). Remove from blotting paper and let it stand for fifteen minutes to allow vapors to evaporate.

---

[33]Opaline is a soft silica powder that cleans the surface of paper.

[34]Blotting paper is highly absorbent, thick paper, designed to absorb ink or other liquids. Acid-free blotting paper is available from archival supply dealers.

⑥ If you have more than one page, interleaving with sheets of silicone release or wax paper before drying under light pressure will prevent sticking of adjacent pages.

METHOD # 2

Many years ago, genealogists learned how to remove the acid from newspaper clippings. They distributed recipes for the process. This outraged librarians and conservators. They claimed it was dangerous and foolish. "If untrained persons attempt deacidification, they may cause irreversible damage to the material and may expose themselves to health hazards. It is, therefore, advisable to engage the services of a professional paper conservator or regional conservation facility." If, on the other hand, you asked them if the recipe the genealogists were using worked, they would say, "Yes, but...."

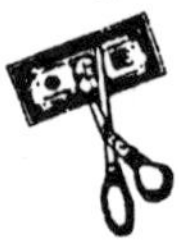

This method can be tricky. The secret here is experience. It is advisable that you experiment on clippings and documents of lesser or no value before risking your valuable items to your inexperience. Once you feel confident, you can continue.

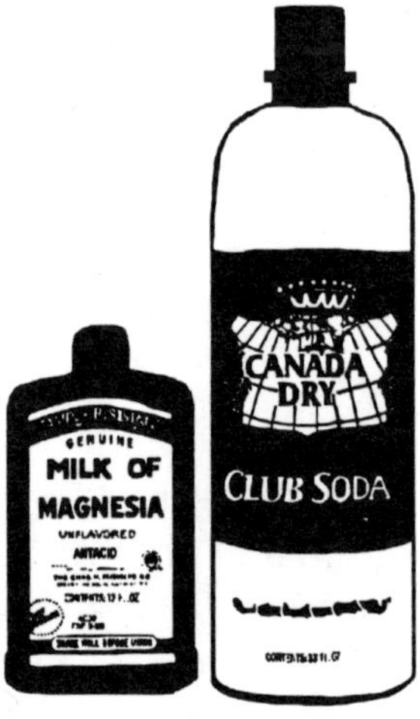

Ingredients for Deacidification Method #2

RECIPE

Materials:

- 1 quart of club soda (unflavored, at room temperature). Any inexpensive brand will do.
- 2 Tablespoons of Milk of Magnesia (unflavored).[35]
- Blotter, paper towels or handi-wipe towels (not pre-moistened mini wipes).
- Pan, large enough to hold the flattened clippings. (A photo tray or clean roasting pan or cake pan (especially glass) work well. It doesn't matter what the pan is made from, the solution is harmless.)
- Nylon (or fiberglass) screen (not metal). (Can be made from an old, but clean window screen or polyester web).[36] Must be large enough to hold the clippings & to grasp the edges of the screen so you don't touch paper.

DIRECTIONS

Dissolve the Milk of Magnesia in the bottle of club soda... **DO NOT SHAKE.** Recap. Gently relieve the pressure once. Very gently, rotate several times.

Let it stand in the refrigerator overnight. Residue will collect at the bottom of the bottle. **Do not mix this residue back into the mix!**

Pour into a pan, being careful not to pour out the residue. Discard bottle and residue.

Test the paper in a non-important place, to be sure inks will not run.

Use the screen under the clipping or document and soak it 5-10 minutes and check continuously to be sure inks do not run.

Continue soaking for one to two hours, remove and pat dry. Use the screen to lift the wet paper or **IT MAY TEAR.**

---

[35]Chemically the magnesium oxide combines with the carbon dioxide of the soda to form magnesium carbonate that neuturalizes the acids of the paper which cause deterioration. Be careful that the brand you choose has no flavorings or medication added. Phillips Milk of Magnesia for example contains undesirable ingredients for this purpose.

[36]You could also use very stiff nylon net or stiff porous fabric.

Wet paper is very weak and susceptible to tearing. Be very cautious in handling it. Use blotters or white paper towels to pat dry. When almost dry, put a book or medium weight onto it to keep it flat, not too heavy or it could affect the design on the document. Keep the weight in place until the paper is completely dry. • **DO NOT REUSE THE SOLUTION! •**

Make a new batch to treat more clippings. You can make two or more bottles at the same time. Consider the amount of space this will require once you lay them out to dry.

If inks do start running after the first 5-10 minutes—remove the paper from the solution—lay it flat and blot carefully to remove moisture. Smeared ink will be picked up by the blotter. Proceed with the drying process as outlined above. Wait several days before you attempt using the non-aqueous solution (method #1).

DISPLAYING KEEPSAKES

Use is very wearing on all paper items and most keepsakes. Properly displayed, they will get handled less than others. It is the oil in a fingerprint that causes damage. To properly display items you need to prevent them from being touched.

At one time it was thought that ultraviolet light caused the fading in color photographs and other items in our homes. But we now know it is all light. An easy rule of thumb to remember is that no matter what kind of keepsake, document or color photographs you have made, they will not last if you expect to display them. (See also page 99). Anytime an item is put on display, even in our own homes, there is a security problem. Whatever is easy to see, can be easily stolen or damaged. Consider making a good color photocopy (even if the original is black and white) to put on display instead of the original. (do not make more than one copy from the original.)

If you display items under glass, be sure there is room for the document (or other item) to be separated from the glass and from the back of the frame or glass case.

- Do clean keepsakes. Remove any surface dirt (see page 37).
- Do cover them or protect them from dust, fingerprints and air pollutants.
- Do display them by placing in frames, in display cases, notebooks or albums.

LAMINATING

Laminating is a process intended to prolong the life of artwork, photos and paper documents. It puts a protective coating on the item and keeps the surface clean from fingerprints and dirt.

Unfortunately most older laminating methods are not safe. Materials used in laminating sheets have been made from cellophane coated with a layer of heat-activated adhesive. Heat is damaging to paper. When you intentionally subject paper to the high temperatures used in laminating, you immediately reduce the life expectancy of paper by at least ten years. The adhesive used on the material is highly acidic and impregnates the paper. The damage is irreversible.

Library supply and archival supply stores now have materials that are safe to use in laminating. They have dry-mount materials that require no heat. These are safe only if the adhesive used is pH neutral or acid-free. Be sure the film and adhesive are both labeled as acid-free. They should be reversible with water. Even though these materials are safe, they should be used to repair pages of books and on the back side of documents only. Avoid laminating anything to the front of a document and never to the front of a photograph.

CONTACT PAPER

Contact paper is made very much like laminating paper, except the adhesive is pressure sensitive, not heat-activated. When this comes in contact with the document, it is impossible to remove without damaging the surface.

Laminating a document or using contact paper on it temporarily keeps marks and stains off, but within just a few years, the acid from the adhesive turns the document or paper yellow. If you were to look at it in a microscope, you could see that the acid has destroyed much of the paper itself. Both procedures are irreversible and destroy the item as a prospective heirloom.

PAGE PROTECTORS

There are many types of page protectors that have been used over the years. One of the first was glassine. It is a thin dense transparent or semi-transparent paper highly resistant to the passage of air and grease. Glassine has been used for the storage of film and negatives for many years until it was found to yellow then brown and crack very easily. It is highly acidic and contains a plasticizer and glycerine. It is dangerous to any item stored in it for even a short time. Some special newer glassine envelopes may be safe.

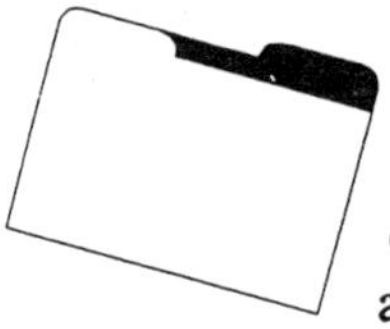

Kraft paper and manila paper are also acidic, but have been used to make folders and envelopes for as long as I can remember. These are dangerous for storing paperwork, film and negatives for any length of time. New acid free folders of all shapes and styles are now available.

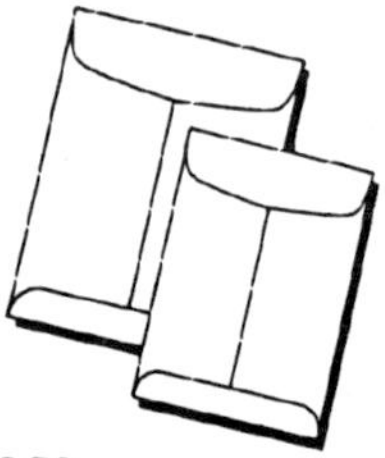

Acrylic is made from acrylic acid and synthetic textile fibers. It may or may not be safe to use as a page protector on paper. Cellophane and acetate (discussed on page 12) should never be used to wrap or protect keepsakes because of their acid content and because they will not last very long.

Plastics are generally made from petroleum products. Two common types are urethane and vinyl. Both contain harsh acids that are released as the material ages. The word vinyl is sometimes used to refer to any of a number of plastics, many are unsafe for preservation.

PolyVinyl chloride (PVC)[37] is cheap and easy to fabricate, but it exudes salts and hydrochloric acid which are very destructive to film emulsion and paper. Vinyl pages may leave an oily residue that cannot be removed. They can wrinkle and stick to film, photos and paper. Plasticizers are acidic, but the damage will first be evident because of the sticking problem. This type of plastic will lift off ink, copy machine and computer print from a page. They also will lift the face off photographs. Vinyl has a distinctive odor, especially as it starts to deteriorate. The rule here is, "If you can smell it, don't use it." Polyurethane has characteristics similar to PVC and should not be used. Don't despair—there are safe page protectors.

For safe storage, use page protectors made from polypropylene, polyethylene or **polyester** film.[38] These materials are recommended for safe storage. Polyester has the highest rating and will last the longest. Polyethylene[39] and polypropylene[40] are acceptable; although not ideal, they will last longer than most other materials. They are heat sensitive and could be damaged if left in direct sunlight or in extreme heat.

Page protectors come in many styles and are used with three-ring binders. Top loaders, where material is inserted from the top, are the most common. Another style is the side opening, where the side nearest the three-rings are open. They are also available for negatives, film, prints and special collections like stamps, baseball cards, etc.

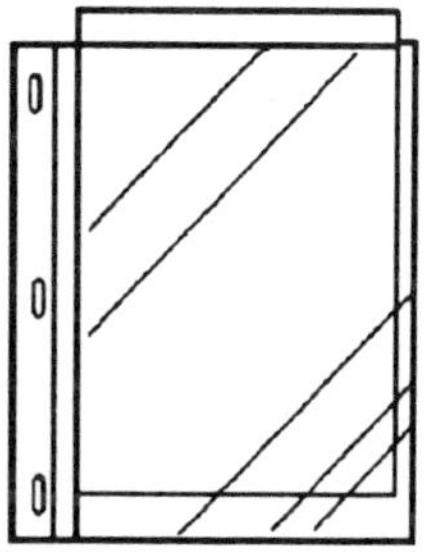

[37]Do not confuse PVA with PVC. One is safe, the other is not. One is a glue the other is sheet plastic.

[38]Polyester is defined as any group of polymers that consist basically of units of ester and are used in making fibers and plastics. It is an inert product and contains nothing that is harmful.

[39]Defined by Webster as *a polymer of ethylene: any of various partially crystalline lightweight thermoplastics that are resistant to chemicals and moisture, have good insulating properties, and are used in packaging and insulations.*

[40]Defined as *Any of various thermoplastics plastics or fibers that are polymers of propylene.* Propylene is a flammable gaseous hydrocarbon obtained by cracking petroleum hydrocarbons and is used chiefly in organic synthesis. Since it is a derivative of petroleum, it is not perfectly archival.

Polyethylene is the material from which better sandwich and food storage bags are made. Page protectors are not usually made from this but paperwork can be safely stored in the food storage size baggies for a limited time. Photographs and film can also be stored in them. Thin polyethylene is clear but when it is thicker, it is not as clear, and items must be removed to see them. Another problem, however, is that this plastic was designed to keep moisture out, but it also keeps moisture in which is good for food, but not for most paperwork. Polyethylene is well suited for film and negatives because there is no static cling and it won't attract dust.

Polypropylene page protectors are available in clear or non-glare. The clear is best. Non-glare means it has a textured surface which makes it hard to see through.[41] The clear type may be more expensive but is worth the price difference. Polypropylene has no static cling. It is **not** recommended for direct contact with photographic film and is not suited for storing slides or negatives. Polypropylene and acetate are highly flammable.

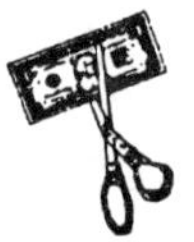

These page protectors can be found at most stationery, department and discount stores and are clearly marked as polypropylene or state they will not lift off type.

- Don't use glassine envelopes on film or negatives.
- Don't use manila folders to store items.
- Don't use page protectors made from cellophane.
- Don't use items made from PVC.
- Don't use polypropylene in direct contact with film.

- Do use polypropylene and polyethylene for economy, but protect from sunlight and heat and make sure items are dry.
- Do use polyester for best storage, protection and encapsulation.

[41] Sometimes the film is clouded or a slight embossing, such as "orange peel" is added. Clarity is always compromised.

ZIP-LOCK BAGS

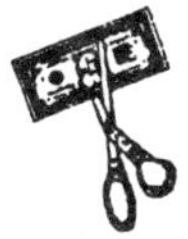

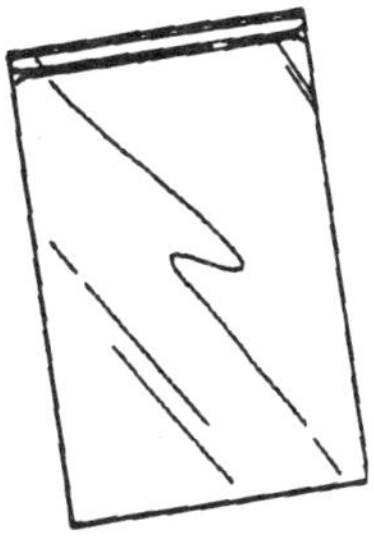

Zip-lock bags are very handy, but regular household types are not acceptable for long term storage or displaying valuables. Most of them are made from polyethylene which offer an excellent means of temporary storage of items. They work especially well for odd-shaped items. These should not be used for permanent display or storage. If there is a lot of moisture present prints can stick to the plastic.

POLYESTER FILM

Polyester is a space-age product. It was first developed for the aero-space industry as the best material for resisting both extreme heat and cold. It can be boiled and frozen without any degradation. It is crystal clear and is the most chemically inert of all sheet plastics. Polyester is very hard and has very sharp edges. Static cling makes it hard to slide film in and out. This static can attract particles of dust. It is more expensive than any other plastic film, but will outlast all others.

Commercially, three companies produce polyester film that fit archival specifications. They are made under the trade-names: Melinex 516, Scotchpar and Mylar "D."

- **Melinex 516**, made by I.C.I. America, Inc., is not as commonly available to the general public as Mylar "D."

- **Scotchpar**, made by 3M Company is the most common film but this product is not available to the public. The commercial version is **Scotchpak**, a product designed for heat-sealing food products and is used in seal-and-boil pouches. They are resistant to temperature extremes. These pouches were designed to be sealed by a heating element (rod) that fuses the film together. This seal is not flat but is quite irregular which is okay for freezing and cooking but not desirable for genealogical items. It is good for protecting three-dimensional items or those irregularly shaped. An example of such an item would be a pressed flower. Archival quality double-coated tape is made from Scotchpar.

- **MYLAR "D"**, made by E. I. duPont de Nemours & Co., Inc. is the most desirable product for use by genealogists and conservators. Mylar D is impervious to heat and cold. It is especially designed for encapsulation and preserving documents. It is available in a multitude of shapes and sizes. You can buy it by the roll, by the sheet, or individual page protectors and envelopes in a myriad of shapes and designs. The preferred thickness for documents is 3 mil. Mylar D this thick is not meant to be creased or folded and should be cut to desired size.

- Don't use Mylar D to store anything susceptible to static electricity.
- Do seal fragile keepsakes that are bulky in a Scotchpak pouch.
- Do use Mylar D to encapsulate flat documents.

ENCAPSULATION

Encapsulation is a method for archivally preserving valuable and fragile documents. It is not just for historic items. Encapsulation is a good, reversible method of protecting many things you want to preserve—documents, birth certificates, awards, newspaper articles, maps, and more. The item is protected from deterioration caused by handling, moisture, contact with acidic materials, and other harmful elements. The process is also reversible, leaving the document in its original state.

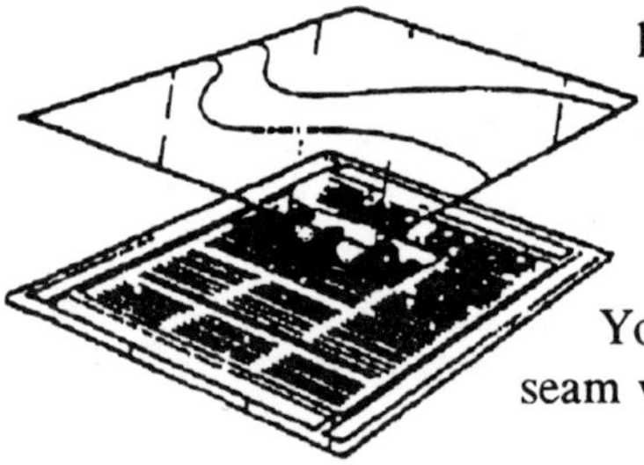

The document is placed between sheets of transparent polyester film (which are larger than the document). You then seal the sheets with ultrasonic seam welding, machine sewing, or a special double-sided film tape made from Scotchpar. Of these three methods, ultrasonic seam welding is impractical for an individual to do because it requires special equipment. If you choose to sew it, use an invisible nylon thread. Start near a corner and use a straight or zigzag sewing stitch. The zigzag looks best. Sew about 1/8 inch out from the document on all sides. End by overlapping stitch about ½ inch. If you make any mistakes, do not reuse the same Mylar "D." The finished look is not as professional looking as using the third method which is the tape.

Pastel or charcoal drawings should never be encapsulated because the static electricity may pick up particles from the page. Photographs and film should never be encapsulated. One danger of encapsulation using double-sided tape is if the document slides to the edge it can possibly get stuck on the tape and the edge of the document could be pulled away. The chances of this happening are slim however.

Before you encapsulate a document, deacidify it according to the directions (on page 33-39). Otherwise, any inherent chemicals may continue to decompose the item.

Supplies for encapsulation may be purchased separately or as kits. The kits tend to be more expensive, but they are convenient for first-time use. On the other hand you may already own some items included in the kit. These supplies may be obtained from an archival supply catalog. The 3M double-sided #415 tape is available from any art supply, library supply or archival supply store (Don't use a substitute tape).

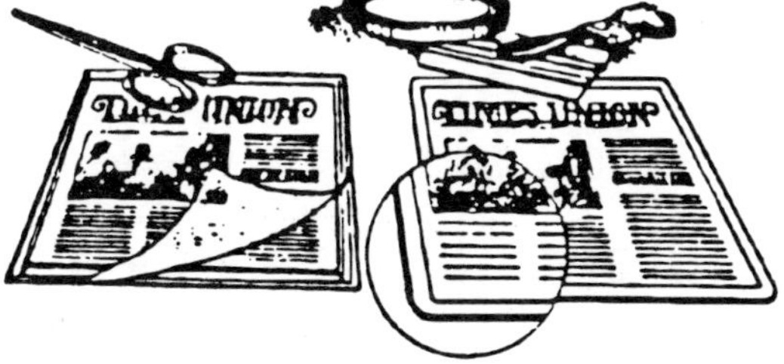

- Do encapsulate fragile or old documents. Do as many as you can afford now and others later.

Materials Required:

- Polyester film (Mylar "D" is preferred) (3 mil)
- Squeegee wiper (like the ones used for cleaning windows)
- Rubber brayer or roller (like a rolling pen with a handle)
- 3M double-sided tape #415 (Scotchpar tape)
- Weight (with clean felt bottom or covered in cloth)
- Clean lint-free cloth (handi-wipe towel)[42]
- Scissors or X-Acto-type knife
- Snippers to round corners (toe nail clippers)

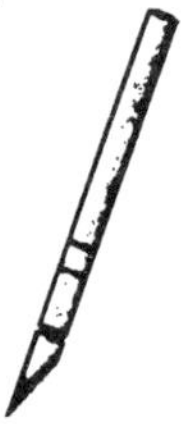

Procedure

➊ Cut two sheets of polyester film at least 1" larger than the document's maximum dimensions on each side.

➋ Place one sheet of the film on a flat, clean, hard work surface. Wipe the entire surface of the film with a lint-free cloth to remove any dust and **create a static charge.**[43]

➌ Place the document in the center of the film making sure that the border is adequate to apply the tape. Place the weight on the center of the document to keep it in place.

➍ Apply the double-faced (double-coated) tape to the bottom piece of film and pull tape very tightly along area to be sealed. (If you don't the polyester will wrinkle.) There should be a 1/8" gap between document and tape. The ends of the tape should be cut square. At least one corner should have a minimum 1/16" gap between tape at corners. (This allows air to escape and the paper to breathe.)

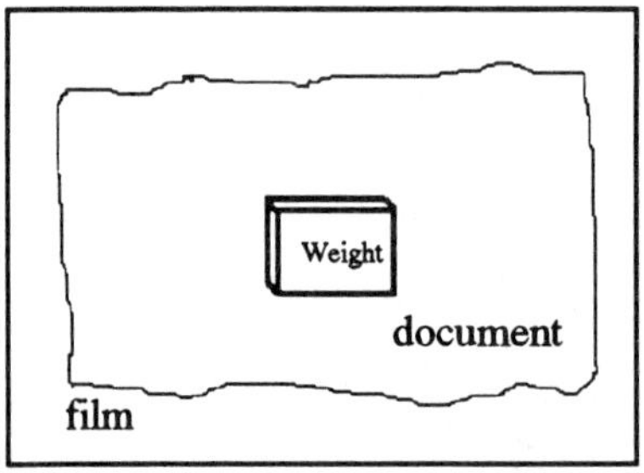

—1 inch margin between edge of polyester and document
—weight in center
—polyester film on top of and below document

[42] Plain or treated dust cloth, any that don't leave harmful residue. They are available from your supermarket or hardware stores. Some brand names include, Handi-wipe, Guardsman, and Stretch 'n dust.

[43] The static charge is important. It helps hold the item to the film and helps the film stick together.

❺ Wipe the second sheet of film with the lint free cloth.
❻ Remove the weight and place the second polyester sheet, cleaned side down, against the document. Square it with the first sheet. (This forms a sandwich, with the document inside).
❼ While holding the partially completed capsule with one hand, use a squeegee and/or brayer to remove as much air as possible from the sandwich. Work from the center to the outside edges. Check to see that the document is in the proper position.
❽ Replace the weight to the center. Lift one corner of the top sheet and carefully peel the protective paper from the tape, working in a counter clockwise direction. Use the squeegee to seal the tape and remove any excess air, working from the center out.
❾ Use the lint free cloth to rub over the areas taped to remove excess air and the brayer to firmly bond the tape to the polyester.
❿ Trim the capsule leaving a slight margin outside the tape on all four sides. A paper cutter gives the straightest edges. Round the corners (use snippers) to prevent scratching or cutting. The polyester is very sharp and will cut both people and other documents unless corners are rounded.

NOTEBOOKS

Inexpensive three-ring binders are commonly made from PVC. In a matter of days or weeks it will destroy documents by lifting the print off the pages. If you use these notebooks, be sure your items are in protective page covers, or encapsulated in polyester. Even then, do not use them for permanent display or storage of items you want to preserve. PVC notebooks can be identified by their odor. Other notebooks are available made of safe materials, but the

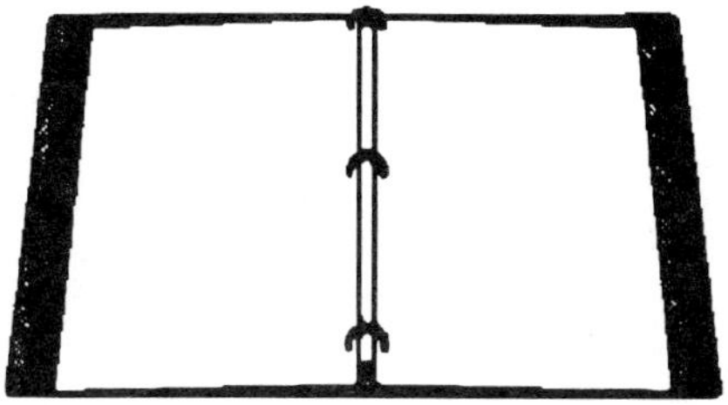

come with D-rings instead of round or O-rings. The D shape allows the pages to lay flat when the notebook is open. The O-ring notebook is easier to turn pages.

- Don't use notebooks or page protectors made from PVC.

- Do use notebooks made from safe material.

ALBUMS & SCRAPBOOKS

Acid-free albums and scrapbooks are available and should be your first choice when purchasing new albums. If you are using old albums, insert acid-free tissue or paper in between pages to protect items. Don't add new items unless they have been encapsulated or inserted in safe page protectors, or at least attached to acid-free paper. (See also page 104)

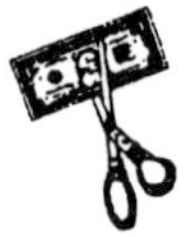

Learn how to arrange album pages in an attractive and pleasing manner. Be creative and imaginative. Use acid-free products. Create an album you will cherish forever! Our paper keepsakes come in many sizes and shapes. The possibilities are virtually unlimited to arrange your valued photos and memorabilia. For information on mounting photos in albums, see page 106.

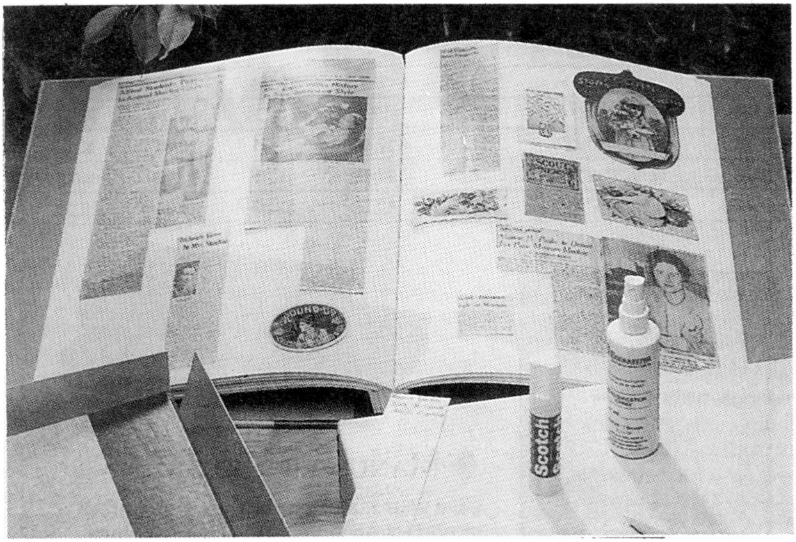

Scrapbook with Newspaper Clippings & Keepsakes

Archival albums usually come with their own slipcases to protect the pages from the environment. Some use D-rings and some use screw posts. The posts allow for easy expansion. These albums come in many sizes and styles. Most of them are quite expensive.

The preferred method for attaching items to album pages is with mounting corners. They should be crystal clear and made of safe products. Mylar D is best. They are available in several styles. Polypropylene corners are also available and are more economical than Mylar D but not quite as high in quality.

- Don't use regular albums found in department stores.
- Do use acid-free albums.
- Do attach items to pages in a safe manner.
- Do use slipcases when possible.

STORING KEEPSAKES

Are your photos and memorabilia stuffed into boxes and drawers? Or worse, are they stored in albums that are chemically destroying the images...and your memories? Is your grandmother's quilt on a shelf in the closet? Where are your albums stored? How long do you want to store them?

Take a few precautions when it comes to storing your keepsakes. Remember they should be clean, laid flat, and placed into appropriate containers. They need to be in a healthy environment, with controlled light, heat and humidity. Keep your own comfort in mind as a gauge. They should be protected from mold and insects.

CLEAN ALL ITEMS TO BE STORED

Make sure your hands are clean, or wear clean white gloves when working with original documents. Inexpensive white cotton gloves sold at photo supply stores are perfect. You can easily see if they get dirty and they can be washed or replaced. It is only when you wear them that you can clearly see just how much dirt there really is on your heirlooms.

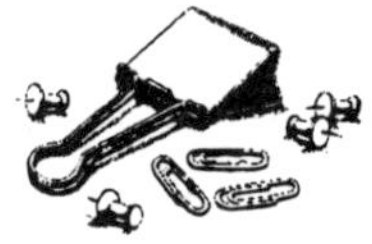

Remove staples, paper clips, rubber bands, or anything attached to the items that is not archival quality. Remove any metal parts or attachments. Staples will rust in a very short time. The rust stain is transferred to the paper and the stain cannot be removed later. Staples made from rust proof Monel steel are available from archival supply companies. The same is true of paper clips. New stainless steel coated clips, or plastic clips are better because they will not rust, but unless the plastic or coating is non-acidic, you may still have a problem.

Rubber bands will deteriorate rapidly. They become a gooey mess, and they quickly lose their elasticity when they are not in active use.

Remove dust, and any other surface dirt or marks. For paper and documents use a soft, clean eraser to remove obvious marks. Do not erase on coated or shiny paper, it will destroy the surface and leave a blemish. It is preferable to use a kneadable pure rubber eraser.[44] This is permanently tacky, pH neutral, and long-lasting. It cleans not only pencil marks, but charcoal, grease, debris and much more. Gum and pink pearl erasers are more abrasive and may tear, wrinkle or leave white spots. Use erasers with caution.

It is also recommended that you use a dry cleaning pad or powder (the type draftsmen use, available at any art supply store). This will help remove the surface dirt, even dirt that is not noticeable to the eye.

Use a photographic emulsion cleaner on photos. It leaves no residue and will not harm the surface of a photo. It removes grease, adhesives, pencil and minor pen marks from both film and prints safely. Practice using it on a less important photo before trying it on one that is valuable. Photo emulsion cleaner is available at most photo supply stores or through archival supply catalogs.

- Do work on important documents with clean hand or white gloves.
- Do remove all foreign objects from documents.
- Do clean surfaces of documents and photos.

KEEP ITEMS FLAT

You should never roll a document if you can avoid it. This is a problem on large documents, certificates, maps, etc. Most libraries store oversized items in large flat metal cabinets. This may not be practical for our homes, but something similar might be considered.

Rolling and folding paper causes the paper fibers to stretch. When an item is rolled the fibers are curved in an unnatural state and after a period of time, the fibers memorize that position and are unable to return to a flat position.

Folding a piece of paper actually puts enough pressure on the fibers to break them in half. Once a page is folded, that becomes the weakest

[44]"Groomstick" is one brand name.

spot on the page. The fibers are not only bent out of shape, but are actually pulled apart. If you were to look at a piece of paper through a magnifying glass, you could see the terrible forces put on the fibers.

Never fold a document if at all possible. Many items in your possession may have already been folded. Letters for example are usually folded to fit in an envelope. Remove them from the envelope and lay them flat. Save the envelopes; they should always stay with the letter.

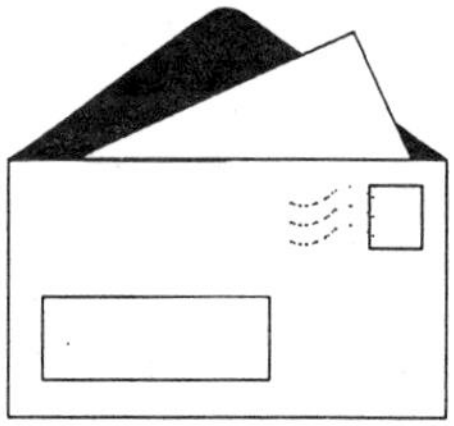

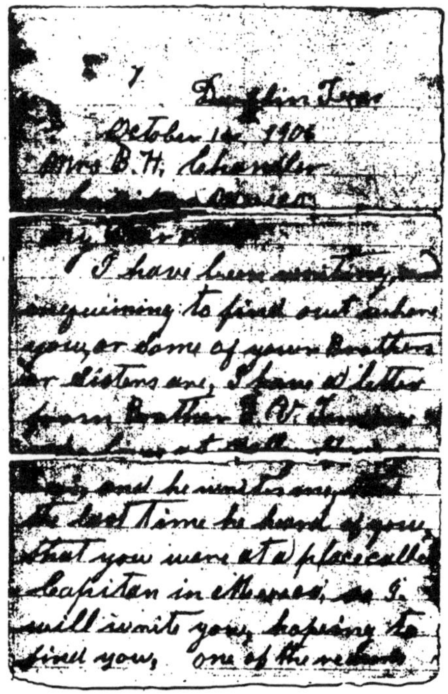

[illegible] Texas
October 14 1906
Mrs B. H. Chandler
[illegible]
[illegible]
I have been waiting and
inquiring to find out where
you or some of your Brothers
or Sisters are, I have a letter
from Brother [illegible]
[illegible]
and he [illegible]
the last time he heard of you
that you were at a place called
Capitan in Illinois, so I
will await you, hoping to
find you, one of the [illegible]

Damage along fold lines
of an old letter

Some people have been known to iron maps and items that have been folded to get them flat enough to store. The heat damages most paper items and cuts down on the item's longevity. There is a better way to flatten paper.

Both rolling and folding are unnatural positions for paper. To preserve items previously stored this way, you need to treat them with a solution (like Wei T'o, see page 35) that will relax the fibers. Before the solution dries, a light, even weight should be applied while it dries completely.

Sometimes several applications are needed to get them completely relaxed so they can lay flat. This is especially true if the items have been this way for a long time.

Another method is to expose the document to high humidity. Humidifying relaxes paper. Allow it to dry under light pressure. This method can take from one-half hour to a full day. You can create your own humidifier by placing a plastic container (large enough to insert the document without rolling or folding) inside a larger plastic container (like a trash can) that has about six inches of warm water at the bottom. Place the document in a pan or container to keep it out of, or above, the water. Cover the large container and wait until the paper becomes limp and damp. Remove it immediately and place on a clean surface. If the document had been rolled, place the inside of the roll down, press between two sheets of polyester web (or clean nylon screen) or silicon release paper making sure there are no wrinkles. If the document had been folded, place it in the same position and press gently on the crease with your finger, press the same as above. If you need to turn the document during treatment, use a microspatula or similar device for carefully handling paper. Allow it to dry under light pressure.

There may be some exceptionally large items that you need to store by rolling or folding. Do it only when it is absolutely necessary. Keep in mind the damage it causes.

- Don't roll or fold documents.
- Do store all paperwork flat (unfolded).

CONTAINERS

Never store in wood or wood products. Wooden file cabinets should never be used. They are attractive to look at, but are murder on paper. You should never store your treasures in a dresser drawer or anything made out of wood, including a cedar chest. For many years we were told to store our heirlooms in a cedar chest to protect them. While it does protect them from mice and insects, it does nothing to protect them from acid in the wood. Bare wood shelves are also dangerous. Since most shelving in our homes is wood, it is best if the wood is painted, varnished, or sealed. Metal shelving is safer.

Never store in cardboard. Cardboard is made from wood products and contains high levels of acid. Use acid-free or archival quality boxes.

- Don't store items in anything made from wood or cardboard.
- Do store items in archival quality boxes and containers.

ARCHIVAL CONTAINERS

Archival quality boxes offer superior protection against the ravages of time and the elements. Archival storage boxes[45] not only have no acid, but contain no lignin. A buffering agent is often added. Check on the buffer used; it may affect colored or dyed items. These boxes are usually made with no adhesives. The corners interlock (are shipped flat) or have metal (baked enamel) corners for extra reinforcement. Archival quality boxes are rigid, durable and expensive. They range in color from a gray or blue-gray with white interiors which have a minimum amount of lignin, to a tan which is lignin free. They are available in many sizes and styles, with flip-top lids or separate lid, with or without a drop-front. Also available are artifact storage boxes (see page 128), these containers are expensive.

[45] Archival boxes are made from purified alpha cellulose pulp board with no sizing.

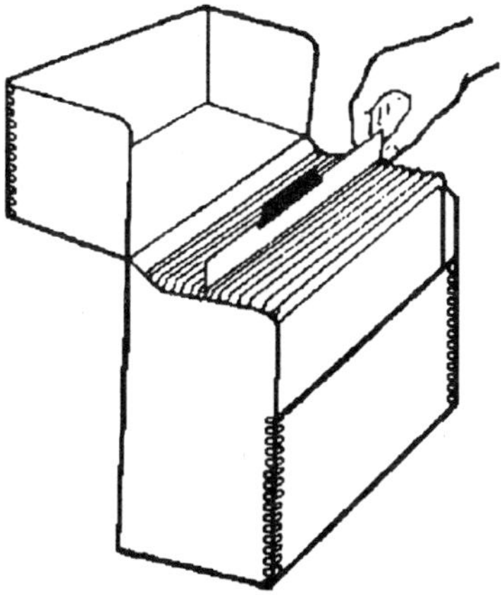

Archival Document Box

When stacking storage boxes of any kind, do not place heavy objects on them or stack them too high. The boxes will break down under the weight and the contents could be crushed or damaged.

- Do clearly mark stored boxes as *Important Papers* or *"Archival—SAVE."*
- Don't stack boxes too high.
- Don't place heavy objects on storage boxes.

BAKED ENAMEL CONTAINERS

Any container made from baked enamel is safe to store your treasures in. Not painted metal but baked enamel. This includes metal file cabinets and portable metal storage boxes. If you cannot afford to purchase the archival boxes you need, put all your valuables in a metal filing cabinet and use acid-free folders, interleaving sheets or tissue, or protect items by encapsulation.

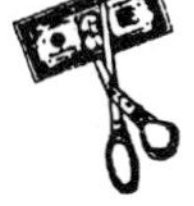

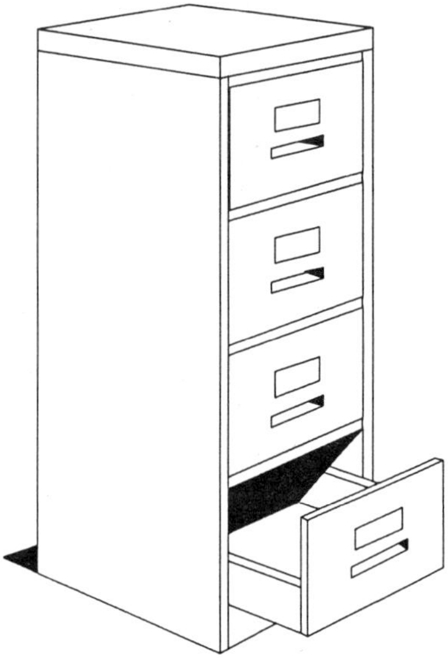

Metal file cabinet for storage

ENVIRONMENTAL CONSIDERATIONS

Environmental considerations when storing items includes temperature, light, humidity, and air circulation. Always store "like-items" together. For instance, all quilts and fabric items should be stored together, all books should be stored together, all paperwork should be stored together, and so forth. This makes it easier to control the environment. Museums control these environmental factors very carefully. There are many devices they use to help. Most of these are not practical for home use, but you may want to check them out (refer to archival supply catalogs).

When you store something, it is best to keep these points in mind: (1) Never store keepsakes, including photos and negatives, in a hot or damp environment. (2) Store them in a cool—dry—dark place with good air circulation. Items should not be down next to the floor, but a few inches above so the air can circulate underneath and around the items stored. Set containers on 2x4 pieces of lumber, bricks, or other support so air can circulate underneath. An interior closet, under the stairs, or in a cupboard, is good. (3) It is best if it is on the inside of your home, not an outside wall. (4) Ideal humidity is between 45-50%. If moisture is present use silica gel or a reusable desiccant canister.[46] This is a small, lightweight container, that can maintain a safe relative humidity in up to three cubic feet of sealed space. It contains a chemically inert material that attracts, condenses, and holds water vapor, and will not harm stored material. (5) All important keepsakes should be stored where there is little or no light. Total darkness is ideal. Make sure any light bulb is low wattage.

- Do store where the heat is fairly constant without extreme fluctuations.
- Do store where the humidity is controlled, where there is good air circulation and where there is little or no light.

MOLD AND FUNGUS

Airborne microorganisms are everywhere and thrive on the surface of photos in humid conditions. It is easy to recognize mold and fungus in our refrigerators or showers. It is a little more difficult to recognize on paper and in photographs. Brown spots on pages of books or documents are not mold but the result of mold. These can be prevented by storing items in a good environment. To control humidity, silica gel should be placed in storage containers. Little packets of silica gel are shipped in boxes of new cameras and electronic devices to reduce moisture. These can be re-used and stored with your collections. Silica gel can also be purchased from a chemical dealer and some archival supply companies.

[46]A desiccant canister is a small (about 3x5 inches) holder that contains a drying agent, such as calcium chloride, and collects moisture. It controls humidity in a small closed area. They are reusable.

HOW TO TREAT MOLD

If it becomes necessary to treat pages that show mold or fungus damage, Use the chemical *"Thymol"* [47] to vapor-treat your documents or books. Thymol comes in crystal form. Make a solution according to bottle directions by mixing water to dissolve the crystals. Saturate a large piece of white towel-paper with the solution. Let it dry completely. Never allow the wet paper to touch the memento. Once the paper is dry, wrap the paper or book with the toweling. Place it in a zip-lock bag. Allow it to set for several days. To retard the fungus from developing you can add a little Thymol solution to mold-sensitive materials like rice or wheat paste (glue). Mix it thoroughly and use as usual. This treatment will stop any new damage but won't remove any existing damage. To prevent future damage, you can place acid-free paper that has been treated (dipped and dryed) in between pages of a book (every nine or ten pages) during storage.

- Don't allow the chemical soaked wet paper to come in contact with documents or books.

INSECTS & VERMIN

Many insects not only eat the paper or fabric of our keepsakes but also leave acidic droppings. Both cause serious damage.

One of the major insects that damage our books, paperwork and fabrics, is the ordinary house moth. Many treatments have been developed over the years. Many insect traps. powders and pellets are available. The old stand-by, "moth balls" works quite well if you do not mind the smell. It is important however, that the product never actually touches the item being stored. The vapors from the balls or crystals kill the insects. Place them in cloth and make sure they never come in direct contact with your treasures.

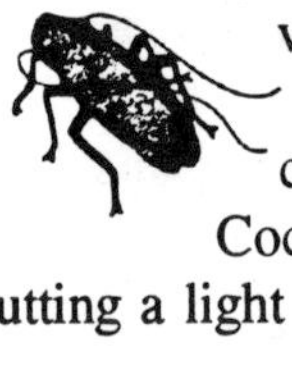

Cockroaches can be stopped from getting into books by putting a light coating of boric acid on shelves behind the books where

[47] Available from archival supply companies. It is a fungicide. It should be used with caution, in a well ventilated area.

children and pets cannot reach it. Fly specks are extremely acidic. Silverfish and spiders can cause a lot of damage especially to sized paper or starched clothing. They can be treated by common insecticides.[48] Do not spray in the air or on the items because the poison leaves oily spots and could harm books and paper.

Bookworms[49] can be killed by freezing the infested item at -20 degrees for seventy-two hours (three days). It should be sealed in a plastic bag (zip-lock) before freezing to prevent moisture from penetrating the paper. After the time is up, allow the moisture on the outside of the bag to evaporate before removing the item. This takes about twelve hours. Freezing will also kill most other insects.
Use commonly recommended methods of extermination to get rid of rodents.

- Don't allow any insecticide or chemical products to come into direct contact with paperwork or keepsakes.

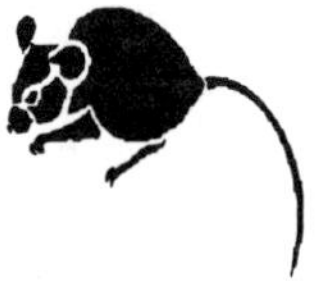

WATER DAMAGE

HOW TO REMOVE WATER STAINS

Books and documents subjected to water damage frequently have water stains. I used to think the stains were from dirty water but have since discovered that is not true. It is the sizing, used in the manufacture of the paper, which becomes concentrated. It can generally be removed by simply soaking the pages in warm water. Test one or two pages before treating a whole book or document. Never attempt to treat coated or clay-filled papers in this way. A smooth or glossy page may indicate it is has a clay filling or is otherwise coated. Treating these papers would not solve the water damage problem and might create new problems.

Consider the value of the books and documents before deciding to

---

[48] Baygon (phenyl Methylcarbomate) or Paradichlorobenzine crystals and others.

[49] Bookworms are the larvae of beetles. They burrow into paper, causing considerable damage. The adult feeds on the paste and binding of the book in an attempt to burrow out to mate.

salvage them. Probably it is less expensive to replace them than to treat them.

To prepare for this treatment, clean off any adhesive and surface dirt before placing them in the water (see page 37), like you would if you were going to store them. Delay any repair work until after the pages are soaked and dried.

Materials Needed:

- Pan - 2 inches (or more) deep & larger in both dimensions than pages
- Paper towels (lint-free cloth)
- Nylon screen, [50] or polyester web
- Pulpboards:[51] clean, larger than page & 1 more than number of pages to be cleaned
- 2 Pressboards[52]
- Brick (covered with clean cloth or clean white sock)
- Warm water (distilled is preferred but plain may do)

Directions

Fill the pan with water to a depth about one inch. Add the pages one at a time. Lower them into water on the screen. Do not overlap the pages. Thoroughly wet each page before adding the next.

Soak for about one hour. At the end of this time, if the water is very discolored, it should be drained off, fresh water added, and soaking continued for another hour.

Carefully remove the pages on the screens one at a time. Lay them on the paper towel (or other white wastepaper) and pat with dry paper towels to remove some of the moisture.

Transfer to a pulpboard, taking care that all wrinkles and folds are

[50] Do not use metallic screen. Cover edges with cloth tape.

[51] Pulpboard is a cream-colored, smooth finished board made from wood pulp. It makes a good working surface to press pages. The weight should be about .045. Or you can use any type of inexpensive, undyed, lightweight board that has a smooth surface. Chipboard will also work. These boards become acidic quickly, so they are only good to temporarily use to flatten paper, not for any type of storage.

[52] Pressboards are rectangular pieces of wood. They can be made from hardwood (plywood if it is smooth enough) or tempered masonite that is smooth on both sides.

flattened out. Cover with another pulpboard and continue with the next page. Repeat the process. When finished, place the stack (pulpboard should be on the top and bottom and in between each page) between the pieces of pressboards and weight it down (just heavy enough to hold it flat). After several hours change the pulpboards and weight it down again until thoroughly dry—overnight or longer. Check for any signs of mold development during drying process. Never use a mechanical press to force water from page. The pages are now ready for any necessary page repair or treatment.

LARGE COLLECTIONS

In case a large number of documents or books are soaked in a flood you should seek the advice and help of book and paper conservators with experience in salvaging water-damaged materials as soon as possible. Take a few steps to prevent further damage by turning off heat and creating free circulation of air. Keep fans and air-conditioning on at night. Keep air circulating to prevent mold growth.

If there are too many books or documents to treat at one time, they can be frozen. Wrap the pages carefully in freezer paper, wax paper or silicone paper as you would fresh meat, and freeze them in a flat position. At some later date you can thaw them out and dry them properly. They can stay frozen up to about six years.

When the Los Angeles Public Library burned, most of the damage to books was caused by water. They freeze-dried the books and salvaged many of them. This method is cheaper than hand drying, but acid migration is a problem during the drying process. It is the preferred method of handling water-logged books and document collections. Freeze dry techniques cause water to pass from the frozen to vapor phase without going to the liquid phase. After the drying process is completed restoration may be needed.

If documents or books are damaged by mud, let them dry completely before trying to remove the mud. Once dry, you can usually brush it off.

- Don't attempt to pick up wet paper unless it is supported because it is very weak and will tear at a touch.

- Don't attempt to treat coated or clay-filled papers.
- Don't attempt to separate single-sheet materials (single pages) unless they are supported on polyester web, stiff fabric or a nylon screen.

- Do check pages as they are drying to be sure there is no mold.
- Do allow mud damaged documents and books to dry before removing mud.

OTHER STAINS

Stains of unknown origin or indeterminate age may or may not be removed. It is a matter of trial and error, and unless done with extreme caution, further damage might occur. It is best to have a professional attempt to remove these stains.

MAINTENANCE

Maintenance of your family keepsakes and important documents must be an on-going process. You must not just store them and forget them. Check on the condition of your collections at least once a year—no longer than two years. If you find any inks or photos are showing signs of fading—make copies before it is too late.

Many points covered in this chapter on documents apply not only to paper but to all of your treasures and heirlooms.

CHAPTER FOUR

BOOKS

Bound books as we know them have been in existence for nearly two thousand years. Over the years, people have said books are the mark of an educated man. Much knowledge and enjoyment is obtained from books. Many people love books not only for their content but for their appearance. There are billions of books in the world. Many are priceless and some are worthless, but most will not last very long because of the way they were made.

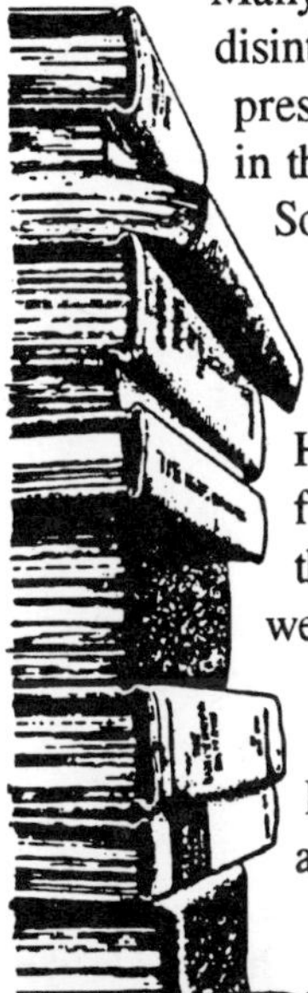

Many books manufactured just twenty years ago are disintegrating. Libraries are having major problems preserving their collections. The quality of the paper used in the manufacture of the book determines its longevity. Soundness of the binding is another consideration.

Most of our homes have books in them. Some have more than others. Some of us only have a few reference books, others may have a library of books. However many books you have, you need to know a few things about them. One of the most important things to know is how to treat them so they will last so we can hand them down to our descendants.

What are your most valuable books? A book of poems? Your favorite novel? A favored Cookbook or How-To book? Do you own any first editions, one-of-a-kind collectors books or books autographed by the author?

My most valuable books include a few out-of-print books, a family Bible, and numerous family histories. I do own books that were more expensive, but even those are not as precious.

PUBLISHING YOUR OWN BOOKS[53]

Publishing a book means to have copies printed and bound for distribution to other people. It can be just a few copies, hundreds or thousands. Not very many people publish their own books, but it is becoming more common all the time because modern technology makes it so simple to do.

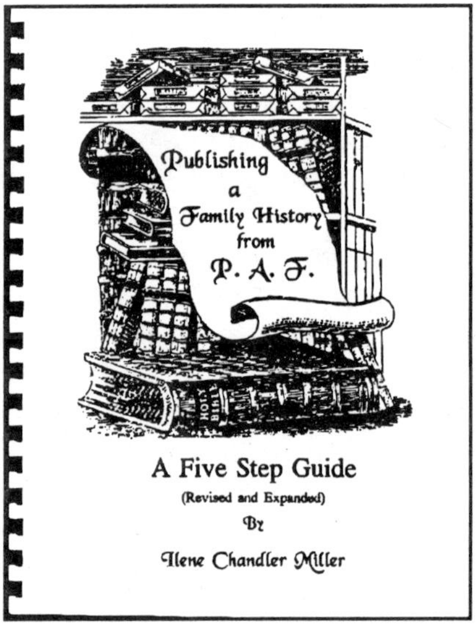

Writing a history about your family and ancestors is easier than ever before because of computers. The modern computer has made it possible for anyone that is interested to publish the results of family research in book form. Publishing your family history is a wonderful way to share information with others. It is the best way to make sure

[53] Ilene (Chandler) Miller, Publishing A Family History from P.A.F. (Personal Ancestral File), (Garden Grove, CA: Shumway Family History Services; 1994), 103 pages.

way to share information with others. It is the best way to make sure not only the present, but future generations know about the past. It is one of the most valuable contributions you can possibly make to your family from those in the past to the yet unborn future generations. It can be as fancy or as plain, as large or small as you decide. You do not need to have very many copies of your book printed unless you want to. Do enough to donate a few to some of the genealogy collections such as the Family History Center in Salt Lake City, Utah and genealogy libraries or historical societies in your area.

When you decide to write a family history, you should select the best quality paper you can possibly afford. You should also select a binding that is durable. This can increase the cost of your book considerably but will be appreciated greatly in years to come.

Your personal journals may be candidates for being bound to give them a better cover. You may want to bind a collection of poems or drawings to give as a special gift, or just to keep as a keepsake. Most binding companies can bind one of a kind or more.

There are many types of binding available, some more durable than others. When having a book bound, a hard bound cover with a rounded back (spine) is the best, most attractive and most durable method of binding. The best cover material is buckram, imitation leather, binder's cloth or real leather.

- Don't procrastinate—no one else can do it as well as you!
- Do publish your own family history.
- Do consider binding special collections.
- Do have them attractively bound for permanence.

EVALUATE TREATMENT OF BOOKS

Carefully evaluate the significance of the book you want to preserve. The importance is up to you to decide whether or not it is worth preserving.

Most of the books in your possession were printed by someone else. Since you have no control on how the book was made or the type of paper it was printed on, treating the paper is your only option. Treating it to remove acid is one way to preserve the book. This can be very expensive and time consuming.

It is impractical and almost foolish to preserve books that will not maintain their value. The books on the shelves of your home may or may not show signs of deterioration. If you do nothing to a book, it will last many years, depending on the type of paper used. How old are the books in your collection? I have many that are more than forty years old. Some of them are in good condition, but many of them are not. Most of them are probably not important enough to treat. Many of them could be replaced if needed. I own few first editions, but I do own some published genealogies that are very important to me and I want my descendants to have them someday.

FUNCTIONAL BOOKS

Any book designed for heavy use could be called a functional book. Using a book is very hard on it.

A child's coloring book will not last very long. It is printed on highly acidic paper and was meant to be used right away, and possibly discarded. Telephone books also fall into this category.

Paperback novels are often printed on this less expensive paper and usually will not last very long. They were not meant to last.

Magazines and books printed on a higher quality paper will last longer, but probably not too much longer before they show signs of yellowing. Textbooks were designed to be used for a limited period of time. There is no need for them to last indefinitely. They become outdated and are usually replaced on a regular basis.

If one of these books becomes damaged, it can be repaired with less caution than a book you want to last. If all you need to do is keep a book from falling apart, you can use any strong tape to do the job. If you want that book to last long enough to hand down to your children or grandchildren, you'd better think twice before using tape of any kind.

- Don't spend a lot of time and money to preserve a book if it is not going to be kept for a long time.

CARING FOR LEATHER BOOKS

Traditionally, leather has been an important covering material for books. All leather is subject to acid penetration from the environment and must be conditioned on a regular basis. Leather becomes very dry as it ages. Every time a book is opened, old leather cracks and flakes off or becomes powdery. To care for leather books they should be periodically waxed or treated with a leather dressing.[54] This will keep the binding soft and flexible. Be careful to keep this solution off the paper inside the book. If the leather is extremely old and dry it will probably have turned a light tan color. Treating it will turn the color dark. Some people think this is an unnatural color for the book, but may actually be the original color of the leather.

- Do treat leather books to keep them in good condition.
- Don't let leather dressing touch paper.

FAMILY BIBLES

Owning a family Bible is a great honor, especially if that Bible has been handed down for several generations and contains family genealogy information. These records are usually written on special pages at the front, center or back of the Bible. Ancestors might have written with pencil or ink. To retard acid migration from the ink or pages, place sheets of acid-free tissue on both sides of the pages having data.

If it is a very old Bible, or has been used a lot, it may need to be repaired. Once a family Bible has been repaired, it should be stored in an acid-free buffered box.

Refer to genealogical publications like the *Genealogical Helper* for ads from companies that do this type of restoration.

[54] Leather dressing is a mixture of lanolin and Neats-foot oil.

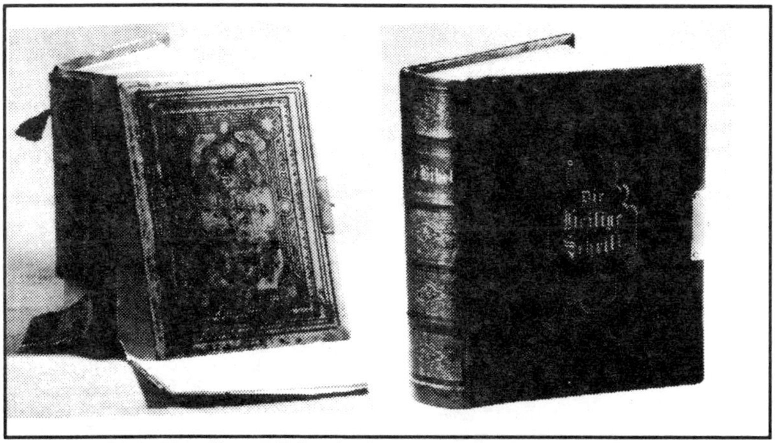

Old Family Bible before restoration and after

RE-BINDING vs RESTORING OLD BOOKS

Rebinding a book means to replace the binding to keep the book intact. The new binding does not necessarily need to be the same as the original, just newer. Books should be rebound when they have loose or detached covers, when the leather is rough or powdery, when most of the pages are acidic and the book must be taken apart to be treated.

Restoring a book means to replace the binding with high quality materials to enhance its beauty and/or value. Restoring a book can also mean treating the pages to remove acid and repairs to the paper as needed. Restoration is done when the leather feels dirty to handle. Good examples of books that should be restored, rather than being rebound are family Bibles, first editions, autographed copies, early manuscript volumes, or a book whose binding is distinctive either for its excellence or as an oddity. Old leather on books should be backed with new leather and the original leather should be rejuvenated. It should then be solidly and soundly bound.

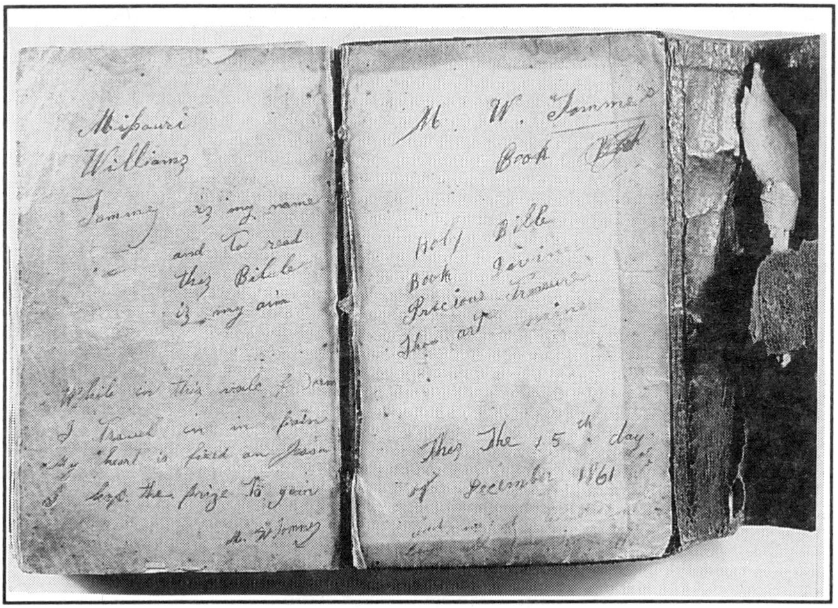

Badly Damaged Family Bible

Quality binding is always getting more expensive. Custom binding of individual books can be especially costly. Of course, if cost is no object, there are craftsmen that specialize in this nearly lost art and do beautiful work. Fine leather covers are gorgeous but require a lot of hand work and expertise. The cost reflects the quality (fine leather bindings cost from $50-200 per book).

Libraries usually rebind books on a regular basis. Library books receive heavy and hard usage. It is usually less expensive to rebind a book than to purchase a replacement copy. Most library staffs are trained in binding and preservation techniques. They usually choose binding that is the least expensive type.

If you cannot afford to rebind or restore the binding on a book immediately, interleave acid-free paper (or tissue) between acidic pages, wrap the volume carefully in acid-free paper, put it in an acid-free container and put it away until some later date.

Some conservators believe that many old books should not be rebound or restored, but preserved in their existing condition in a protective case. This would assume that their projected use is nominal and that they are not in a state of deterioration.

DEACIDIFYING PAGES OF A BOOK

Once it is decided that the pages of a book contain high levels of acid, you may need to remove the acid. Most family Bibles were printed on good paper and probably will not need to be treated.

The easiest treatment is to spray them with Wei T'o spray, using blotting paper between the pages until they dry. Do only a few pages at a time to allow them to dry thoroughly before proceeding. (See procedure on page 37)

This can be a very time-consuming process. It is also an expensive process. Wei T'o spray costs over $23 for a 22 oz. can (unless purchased in quantity). One can will treat about twenty-five 8½x11 pages or more smaller pages, depending on the amount of spray you use. As a rule of thumb, estimate about $1 per page (that does both sides of the page). A 6x9" page will require almost half that amount.

WATER DAMAGED BOOKS

If a whole book is damaged, it may have to be taken apart to be properly treated. This is best done by a professional. The book would then need to be rebound, using the original cover material if it is in good condition, or it might need to be replaced. Consider the value of the book before attempting to salvage any that are seriously water damaged. (See also water damage on page 61).

Books that have been seriously water damaged need immediate treatment to prevent mold from developing. They need to first be washed with clean cool running water. This must be done gently just to remove loose mud or debris. Use a sponge with a dabbing motion; do not rub. All of the mud will not come off. It is easiest to remove this mud after the book is dry. Hold a book firmly when it is being washed. A closed book is resistant to further damage. Do not remove the covers from books, as they help to support the books during drying. Clean, white blotter paper, white paper towels and unprinted newsprint may be used for interleaving in the drying process. When partially dry, books may be hung over nylon lines to finish drying (wet books are too heavy and hanging them will cause damage). The binding will be the last part of the book to dry. During the drying process, check the inside of the book often for signs of mold development.[55] Treat if necessary (see page 60).

- Don't open a wet book.
- Don't remove covers from wet books.
- Don't scrub or rub a wet book.
- Do seek professional advice as soon as possible.

[55] Waters, Peter, *Procedures for Salvage of Water-Damaged Materials.* (Washington DC: Library of Congress, 1975.)

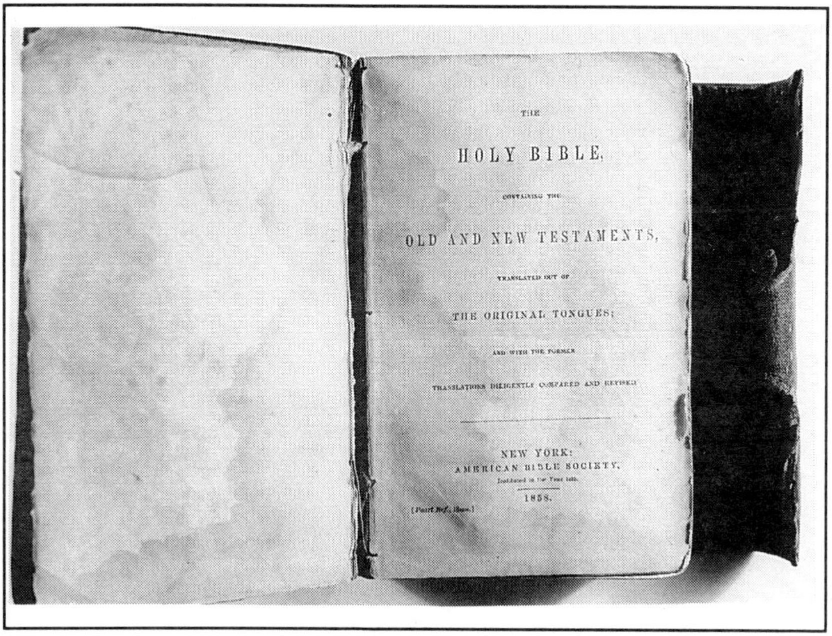

Water Damaged Bible

BOOK REPAIRS

Books are constantly abused and suffer damage including torn or wrinkled pages, ripped or stained covers. Some repairs are easy and can be quickly made. Wrinkled or folded pages can be straightened out by getting them slightly damp (humid), then inserting wax paper on both sides and pressing closed till dry. Torn pages can be mended using the same techniques covered for paper (see page 26). The same is true for cleaning pages (see page 37).

A simple, soft eraser will remove many smudges. A dry-cleaning pad also works well on the pages. Sometimes ink and ballpoint pen can be removed with household bleach and a tooth pick. Use extreme caution, bleach is caustic and can damage the paper. This will not work on all ink. Another solution you might try is oxalic acid crystals (dissolve a few in a small amount of water) and use the tooth pick to dab it onto the page. This chemical is a poisonous strong acid. It is not recommended for use on valuable paper or books.

Some book covers can be cleaned easily with a soft damp cloth and a small amount of mild detergent, unless the cover is sized with starch. Test one corner to be sure the cover material does not contain starch which water will damage.[56] A small dab of glue (especially PVA) might help repair loose end sheets and covers. Use glue cautiously. Especially valuable books should be repaired with great caution, repairs may lower the value of the book. Better to do nothing than to do the wrong thing!

COVERS AND SLIPCASES

Protective covers for books protect them from dust; that is why a cover is usually called a dust jacket. This should not be removed from books. If the book is going to be handled a lot, you should consider covering the dust jacket with plastic (mylar or polypropylene).

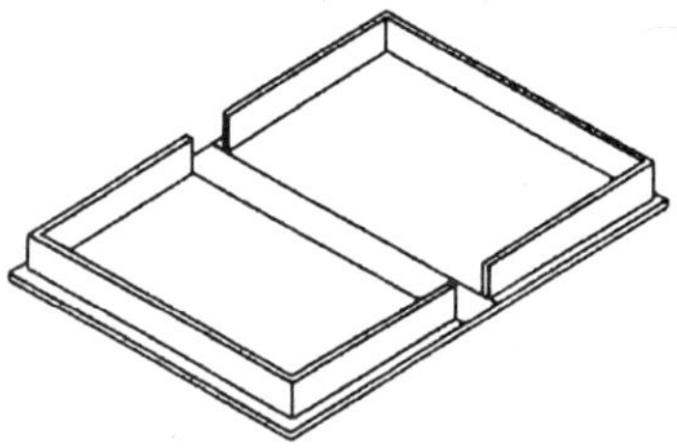

Clamshell Style Box

Protective covers especially designed for books include clamshell boxes. They are one piece boxes that completely cover the book.

A slipcase is a container specifically made for a book that fits perfectly around the book, in which the book slips in and out very easily. Slipcases protect both the book and the contents. It is an appropriate way to store books on a shelf because it keeps pollutants in the air from penetrating the edges of a book. Although the spine of the book is not protected, you can easily see the title of the book while it is on a shelf. Do not store leather books in slipcases unless they are wrapped in a special covering first.[57]

Occasionally a slip case that fits too tightly may cause a problem.

[56] If it does contain starch, it will become soft, it could become limp, it might ooze and change color.

[57] This special covering is called a chemise and is usually supplied with the slipcase designed specifically for leather books.

A slight vacuum could be created, making it hard to remove the book. If you force the book, it might be damaged. To alleviate this possibility put a tiny hole into the back or bottom of the slipcase (any sharp object can be used).

Most fine books come with their own slipcase, including expensive art books. The slipcase adds a lot to the cost of a book, but will add years to its life. They are usually covered in a material matching or complimenting the book.

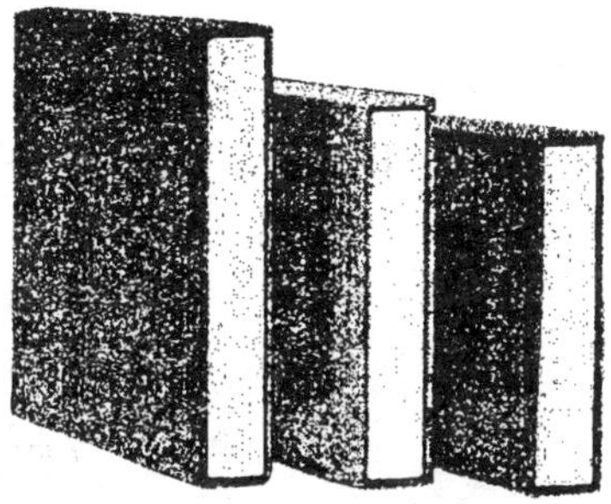

Slipcases

You can make your own slipcase with a little bit of ingenuity and acid free supplies. Pay special attention to the materials selected to line the case. They should be acid-free and free of any decoration that might transfer to the book.

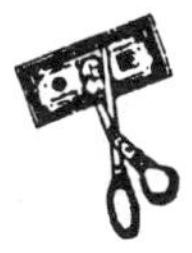

When you are considering buying new books or albums, choose those that already have slipcases.

STORING BOOKS

Books should be stored in an upright position on a book shelf. A shelf is safest when it is sealed or painted to cover bare wood. Baked enamel shelving (metal shelves) are much safer. Books were designed to be sitting in an upright position securely supported on both sides. They should not be supported too tightly or too loosely. If they lean, the pages of the book are prone to becoming bent or pulled out of the binding. When the book is tilted the spine is under great strain and the book can be seriously damaged. Oversized books should lay flat. Remove a book from the shelf by grasping the middle of the spine.

If books are to be stored on shelves in open areas, consider using some type of protective case; if not a slipcase, use a Solander case or folding box. These are basically double boxes and look like a closed book. They protect the book on all sides, including the spine. Fireproof cases should be considered for long term storage. Archival boxes are another choice.

Shelved Books

If none of these are available to you, at least place a sheet of acid-free heavy weight paper on top of several books on a shelf. This will protect the books from dust and some pollutants. Periodically replace this paper when it gets dirty.

Books should be dusted regularly. The closer the shelves are together, the fewer air pollutants can get to the books. Even fly specks are acidic and cause serious damage. Allow enough room for air to circulate.

- Don't store books on the floor, in a damp basement or hot attic.

CHAPTER FIVE

PHOTOGRAPHS

Photography was first introduced in 1839. No one had any idea of the impact it would have on future generations. Since that time photographs have been an important part of our daily lives. "A picture is worth a thousand words." They document our lives in a way nothing else can. Old photographs offer the past to us in a way that words cannot, and yet photographs are by far our most perishable resource. It is important that these captured images are saved for posterity.

BLACK AND WHITE VS. COLOR

Black and white photographs, when processed and stored properly, can be archival. This means that they will last at least 100 years and are extremely fade-resistant.[58]

The most popular method today to record events is with color photographs.[59] The life expectancy of a color photograph is from three to twenty-five years. After as little as three years, the color can start to fade. There is **no known way to chemically restore the dye image once it has deteriorated**. Even color photographs kept in dark storage are subject to fading and staining.

---

[58]The reason black and white photos out-last color is primarily due to the fact that in black and white silver is used to obtain the image. Because silver is a metal, it is impervious to many outside influences. However, metals can be changed into another element that can be deteriorated over time. Silver can be affected by coming into contact with anything containing sulfur.

[59]Color photography utilizes three layers of organic dyes instead of one layer of metallic silver to produce an image. Organic dyes are very susceptible to deterioration, because of the destructive nature of light, heat and humidity. Deterioration occurs when there is an unequal fading of one or two of the layers. The three dye layers almost never fade at the same degree or rate.

Photographs require constant care—they begin to chemically self-destruct as soon as they are made. Fading is the most prevalent form of deterioration. Temperature and humidity greatly accelerate fading in both black and white and color photography. Discoloration of a dye in the print is usually what causes yellow staining of color prints.

We used to believe the main cause of degradation in color photographs was ultraviolet (UV) radiation from the sun. Today most modern photo labs incorporate an ultraviolet treatment to protect against UV fading. However, visible light is the primary cause of images fading, according to recent findings. That means any time you expose a color photo to the light, it will start to fade. Constant exposure to light emitted from a 75 watt tungsten bulb is enough to cause noticeable deterioration to photos in as little as four years.

The second most common form of deterioration is called sulfiding or mirroring. This form of deterioration is identified by its shiny, bluish silver surface.[60] This condition usually progresses from the outside edges of the print toward the center. It is also noticeable in the darkest parts of the print. Never touch the surface of a mirrored print, because finger oils will cause changes in the surface.

There is no comparison in the life expectancy of black and white prints versus color. If you have photos that are important to you, be sure you have a black and white copy made before the original photo fades.

We are faced with the dilemma of choosing between color for aesthetically pleasing temporary photographs or black and white to record historically important events. When photographing a wedding, birth, or anniversary, for example, you may want to take black and white and color as well to satisfy both needs.

OTHER DAMAGE

Over the years, through handling by dozens of hands, photographs can be damaged. Finger prints are oily and acidic and cannot help but

[60] It occurs when silver ions migrate to the surface of the gelatin layer and are then reduced back to the metallic state.

damage the image on a negative or photo. They are also damaged by being bent, folded or scratched.

Another type of damage to a photograph happens when we write on the back of the print. The sharp point of the pen or pencil might make a permanent impression on the paper. If you have written on the backs of photo prints with ink and stack them, the ink could easily transfer to the face of the photo underneath. Not because the ink was wet, but because of the type of ink and the paper the photo was printed on. This type of transferred ink can sometimes be removed if treated soon after it happens. Use a good photo emulsion cleaner.

COPY OLD PHOTOS

Copy old family photographs when the originals first show signs of deterioration or before they fade. Photos are often copied when the original is too fragile to handle. Store these old photos instead of displaying them. As a safety precaution, give copies of old photos as gifts to friends and relatives. They are often copied to be sent to someone who is compiling a family history. These are reasons why it is important to discuss copying and the different available methods.

Make a photographic copy of all important images. It insures preservation of that image. It allows you to mail copies and share with others. Copy techniques can restore or enhance tone, contrast, detail or original color. Major stains, tears, and damage can be retouched on the copy. Recently developed technology for scanning photos also provides the ability to enhance contrast, remedies stains and tears, and even corrects other problems in the original photo. While this process is intricate and time consuming, it is becoming easier and better (see page 92).

Consider future use for historical photographs. Historical societies, museums, or collectors may have an interest. Maybe it would be better to give an original photograph of historical significance to the local historical society and just keep a good copy for the family.

Once it is determined why you want copies of the photos, let that be your guide as to the type of copy you will have done.

The first thing to understand is that the main purpose of copying old photographs is to faithfully duplicate the original image. Next, it is to make a new copy print that will last as long, if not longer, than the original.

If you cannot afford to have your old photos copied all at once, do a few at a time. First, start with those that are in the worst condition. Consider having new negatives made, but hold off on having prints made until you can afford it. At least this way you will have good negatives in case your originals fade even more.

Some old documents can be photographically copied to enhance faded writing, ink, as well as some badly faded photos. This process involves the use of polarized and infrared photographic techniques. This technology was previously only used by law enforcement and military applications. It can now be used by anyone using special film. Infrared photography brings out faded inks and pencil writing that has faded.

FILM & NEGATIVES

First you will need a good negative made from the original print. You can make the new negative yourself or have someone else do it.

All photographic prints are made from traditional negatives produced from film. The most common type of film on the market is color film. The colors in these films have been very unstable. As a color negative ages, the colors can change. The next time you have a print made from it, the results may be shocking. What used to be blue could have changed to orange or red, or even gray or green. Newer color films are more stable, but still leave much to be desired.

Black and white film is much harder to find because it is not as popular as it was years ago. Not many people use it anymore. To find it, you may have to go to a specialty camera store. There are many different types and speeds of film. Some are more stable than others. Many books are available that describe the differences in film.

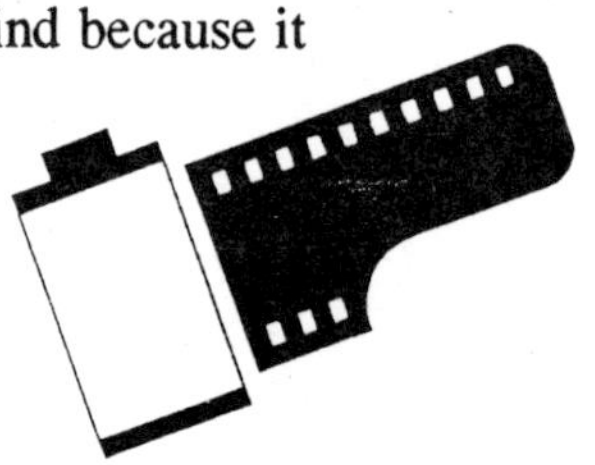

Almost any photographic studio or film processing company can make the copy negative for you. The cost will vary from a couple of dollars each to about fifteen dollars per negative.

I have been doing copy work for many years and my rates are very reasonable.[61] Black and white negatives are about $2 each, with a minimum of four per customer (this is one strip). The reason for the minimum is that single 35mm negatives are hard to handle.[62] Many photo developers will not even handle single negatives because their equipment is set to handle the strips. They may charge you extra for special handling. Other sizes of film are not nearly as common as 35mm and some processing companies do not handle them.

A 4x5 inch negative will give the best results for enlargement purposes. These must be made with special studio-type cameras. They are considerably more expensive to make and sometimes to get printed.

You should not throw away a negative, even if it gets old. That is, unless it is forty-five years old or more (see page 115). As long as you have a negative, you can always make an original print. To get an interesting antique look without the need of having the print toned, copy a black and white photo with **color film**. It would have the same life expectancy as a color print, but makes a good copy for display purposes. (See also page 86).

- Don't cut negatives apart.
- Don't ever throw away a negative.
- Do copy a black and white photo with **color film** to get an interesting effect.

[61]Keepsake Photos, 9712 Mirage Circle, Garden Grove, CA, 92644.

[62]Single 35mm negatives are about an inch in size. It is very hard to grasp them without getting fingerprints on the film. It is likewise hard to put them into an enlarger to be printed. Photo labs frequently attach a strip of paper so they can manipulate the negatives. It is always best to keep 35mm negatives in strips of at least four.

PHOTOGRAPHIC PRINTS

An original print means the one-of-a-kind irreplaceable print made with materials, chemicals and techniques of the same time period as the negative from which it originated.

When a print is made from a negative, it can duplicate the originalimage "as is" or can be photographically manipulated to improve or enhance it. It can be selectively cropped, contrast can be improved, and it can be toned, or the image can be hand-colored. You should consider the cost when deciding the type of processing you will use. The most common printing method includes printing them on Resin-Coated (RC) paper. This paper contains two layers of plastic and very little real paper. The plastic is unstable and will not last as long as 100% paper. RC paper is the kind used by most photo processing companies, including the "One Hour" developers and your local drug store. This is the least expensive method for copying old prints.

This modern type of print (RC) gives you an inexpensive copy of your original prints. They are not archival, but are easy to reprint periodically if you see any signs of deterioration. If you have a good black and white negative, you can always have more copies printed later with the more expensive paper.

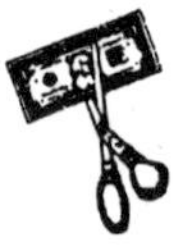

Another choice involves using color film and paper. It will duplicate the "colors and appearance" of the original whether it is in good or bad condition. It is usually not possible to manipulate the process to improve or enhance a deteriorating original image. What is most important, color prints will not maintain their color and will eventually fade. This has been a big problem with old color photographs and new ones.

Don't discard old pictures of yourself just because you don't like them. I have a relative who cuts her face out of every family photo. Someday her descendants are going to wonder what she looked like.

When you make copy prints of your photos, have them printed on glossy paper when you want to have them scanned, make laser copies, have them half-toned, or have a copy negative made.

ARCHIVAL PRINTS

The prints from the local drugstore or One-hour photo developer will never be archival. The best method would be a print that is custom processed for maximum permanence.

It is the paper that the print is on that determines whether or not it is archival. Prints made on 100% double-weight paper and processed properly become archival. Chemicals used in photo processing must be fresh and all prints must be thoroughly washed to remove residual chemicals. Special solutions can be added to clarify the rinse and make sure the print comes out clean. Make the print with a larger than usual border (one inch) around the image. This white space acts as a buffer zone between the image and the edge of the paper, which can be easily contaminated.

The above process is not new. It is the old tried and true method. Before the arrival of plastics, most photos were printed on real paper. If you have a home darkroom and can process your own prints, you can easily produce archival quality prints.

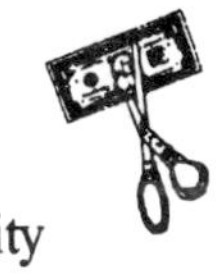

TONING & TINTING PRINTS

Toning a photo changes the color from black and white to sepia, a brown tone or selenium. Sepia is the yellowish-brown tone that makes the photo look old. Brown tone is a more sophisticated look, but also adds character to a restored photo. Selenium or gold toning make the silver in a black and white print more stable.

A black and white print is re-processed and new toner is added. These new colors give the print an antique or old look. These processes are very stable and will not fade like color photos. The cost of toning starts at $10 per 5x7 print and the price goes up from there. This is over and above the cost of printing the black and white print.

For an economical antique look in a print, make a color negative of a black and white print. It gives an interesting look to the copy print, but remember this color print is not stable or permanent. (See also page 83).

RETOUCHING PRINTS

Retouching photographic prints is done to correct flaws, remove blemishes, open closed eyes, soften wrinkles, adjust hair styles, etc. Before we had color photography, if you wanted a color portrait, an artist added the color by hand. This was done either by tinting or painting with oil or water colors. Using this approach, flaws could be covered up or corrected.

Removing small blemishes on a print is not too difficult a task and can make a surprisingly big improvement to the appearance. Professional retouching by hand (usually using an air brush) is expensive and requires the talent of specialists in this field. It is well out of the realm of the average person to do. Retouching by hand requires someone who is familiar not only with photography, but who is also an artist.

We have all seen the results of professionally retouched portraits. Many portraits are retouched by the photographic studio before you even see the finished prints.

RESTORED PHOTOGRAPHS

Usually when we say a photo has been restored, we mean a photo that has sustained damage is repaired. The repair work is not done to the actual, original **damaged** photo. Retouching the new photo after it is copied leaves the damaged photo unaltered. Most people think the actual original is fixed which is not the case.

Restoring a photo involves returning it to its original condition, as much as possible. Repair photos when the photo has been torn, the paper it was printed on has been damaged, or the surface image has been affected in some way. Do it when you want shadowed and unclear photos to be reconstructed to produce clearer images. Restoring does not mean altering it from what it was originally. However, if the photo has sustained major damage (no matter what cause) it is sometimes up to the person doing the restoration to re-create the images. This is where artistic ability is important.

Frequently a restored photo does not look like the original and people may not even look like themselves. But it is often the best that can be done. It is extremely difficult to create something in a photograph from nothing. A seriously damaged photo may resemble an oil painting when it is finished. It becomes an artist's conception of how the original may have looked. Faces are especially difficult to reconstruct. This restoration is much better than not having a usable photo, but it should be identified on the back as having been retouched.

I have a picture of my second great-grandmother that was restored more than forty years ago. For a long time I did not realize it had been retouched. She looked very stately and stern. She died long before I was born. I used to imagine what kind of person she was. My impression from the photo was that she was harsh and rugged. A few years ago a distant relative showed me an unaltered copy of the same photo. She was stately but not nearly as stern looking. In fact, she had a very sweet expression on her face. My whole attitude about her changed.

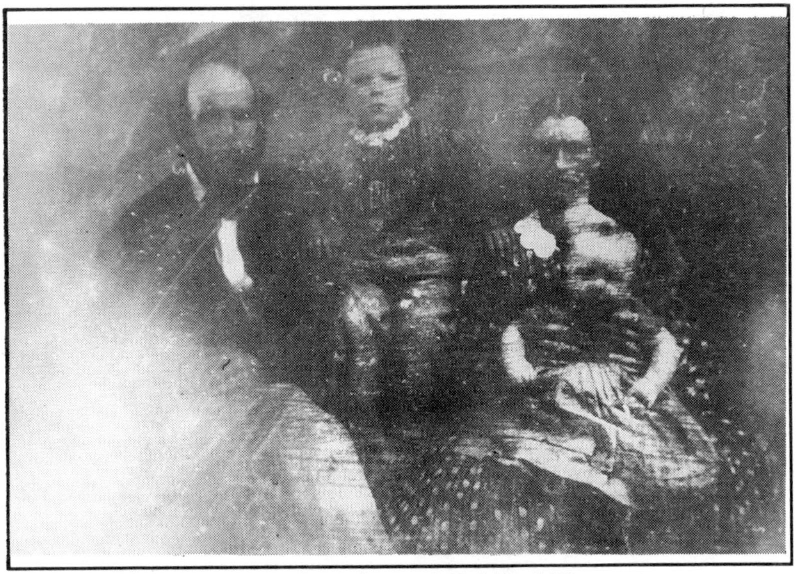

Seriously Deteriorated Original Photo

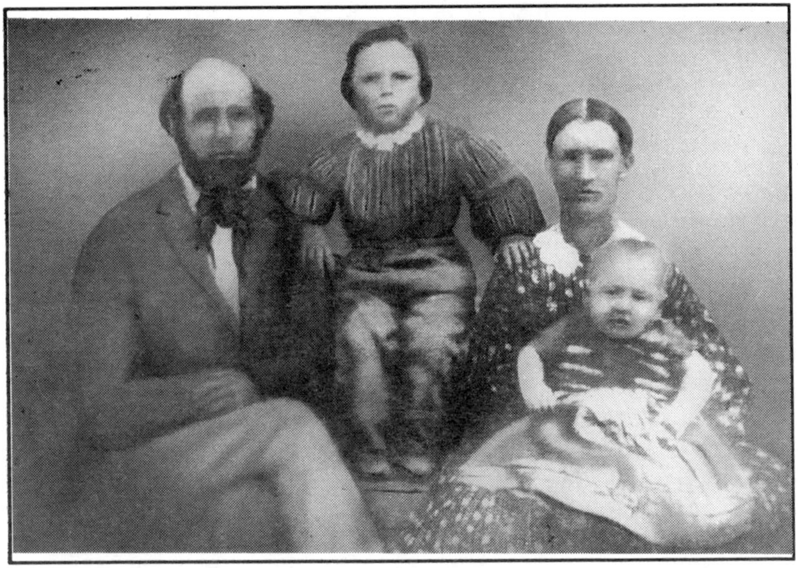

Photo Restored by Artist
Images have been reconstructed. This is an *artist's impression* and does not always look like the original.

LASER COLOR COPIES

A new method of copying photos is to make laser color copies. Use this method to copy black and white and color photos. The black and white copy takes on an antique look. This process produces a copy that is surprisingly close to the original image, with very little loss in detail.

The copier exposes the original photo to a very intense light. Making multiple copies of the original can be harmful. Limit the number of copies made. Instead of copying the original, make multiple copies of a copy. The resulting copy is on a high-quality, but very thin piece of paper. Tearing or damaging this paper can be easily done. This process allows you to make enlargements or reductions in size but no image improvement. Wrinkling or bending the paper can easily crack the color emulsions. It is not known how long these copies will last. Techniques and processes are always improving and may prove to be quite satisfactory for many uses. The resulting print is often near-original quality.

When sharing copies of old photos with anyone who is going to publish a family history, send a good laser color copy, but be sure to mail it flat. This is the preferred method, second only to a copy printed on photographic paper (instead of the original).

I have borrowed lots of photos from people to include in my printed family histories. I have many color copies from which I have made half-tones that turn out beautifully. Halftoning is a process of copying a photograph for publication in a book. It ensures the highest possible reproduction of the photo.

Color laser copies are a wonderful product of new technology. There is a nation-wide copy company (Kinko's) that offers good color copies for reasonable prices (usually less than $2 each). By shopping around, you can save money. Any company that has a Cannon color copier can make them. I have heard by comparison shopping, color copies can be purchased for as little as thirty-nine cents each, especially if you have a lot of them to do at one time or do multiple copies of the same original (which should actually be a copy).

- Don't fold a color copy. It will permanently damage the image.
- Do make laser color copies of old black & white photos.

COMPUTER GENERATED COPIES

The newest method for copying old photos is scanning them on a computer scanner and printing them out on a high quality printer. This is called **Photo Digital Imaging**. The results of this method vary a lot. The biggest advantage of using this method is the ability to retouch, crop and enhance the original image. It allows you to restore the photo to its original or better than original condition. While the computer is a powerful tool, it cannot reproduce something that is not there.

There are many other possibilities of things that can be done in retouching a photo by using a scanner. In a color photo, "red-eye" can be reduced and eyes can be opened. Backgrounds can be changed or improved and flaws removed. Any photo damage can be repaired. Items can be removed, people can be moved or added from one photo to another to create wonderful composites, and much more. (See photos on page 93)

DO-IT-YOURSELF DIGITIZING

If you want to do your own scanning, there are several things to consider. Hand scanners and ordinary printers will not give high quality results. An expensive photo enhancing software program is essential to capture the detail of a fine photo. A scanner and printer that can print in at least 600 DPI (dots per inch) are also necessary. Even then, the best quality you can print out at home is on thin paper. It is beautiful, high quality paper, but its thickness might be a concern. (The problem is the same as with laser color copies).

All this digitizing can be done if you take the time to learn the techniques. You do not have to be an artist or know anything about photography. You do need experience using the computer software and have access to the 600 DPI scanner and printer.

Resulting files can be extremely large. To copy or save the files from your hard drive to disk, do it by using your DOS backup command, or "stack" the disk. The file format that will stack the easiest with the least amount of compression loss is .TIF graphic files.

ECONOMICAL DIGITIZED PHOTOS

There are some mail-order photo companies that can put your photos on disk for much less, but the quality is also much lower.

New advertisements by genealogical publishing and photography companies offer to digitize your photos, then convert them into computer files. Usually they advertise that this can be done for just a few dollars (often about $10) for up to twenty-four photos. This is misleading. Thumbnails are what they are offering. Thumbnails are scanned images that are very small, usually about 1½x2 inches. That is the only way they fit so many on a disk. They are scanned at a low resolution. These photos (especially color) will look fine on your computer monitor. If you print them at this small size, they will also look okay, but once you try to enlarge them, the results are not good. These inexpensive prices never include re-touching and probably not clean-up of minor blemishes. It is in the retouching that a company makes their profit.

PROFESSIONAL SCANNING

There are some photo labs and computer specialists that can scan the information onto a computer disk for you. This can be done with or without retouching. Your scanned pictures are saved to disk in a compressed file format for you to load and manipulate in your photo program. Scanning these images at high resolution will result in large files. A typical 3½x5 print at 600 DPI can give you a 15MB file. This service will cost about $12 per image.

Scanning the image at 200 DPI is less expensive and more images will fit on a disk. Once the image is compressed, a disk will hold about twenty photos. This will cost about $5 per image, and there is often a three-image minimum. The resulting photo will not be as good a quality or give the detail as the one done at 600 DPI (or higher).

Be sure the file format is compatible with your software program, or you have the necessary files from the program that generated the image to disk. I can personally and enthusiastically recommend a company that can do a beautiful job on your digital image retouching. The cost is much more economical than those quoted above, and the quality is wonderful. Contact: Keepsake Photos, 9712 Mirage Circle, Garden Grove, CA 92644

DIGITAL RESTORATION SERVICES

Anything that a professional photo retoucher can do by hand can be done on a computer equipped with special software and hardware. In addition, you can also construct a family group portrait from the best poses in several different photographs. You can separate images and print them individually. Images can be flipped for composition purposes and much more.

Many photo laboratories offer this photo digital image retouching service. High quality work is quite expensive. They charge according to the amount of time it will take to do the necessary restorations. You can expect to pay approximately $75 per hour in most areas of the country. It can take two hours or more to restore a badly damaged original, or move people around.

Minor retouching, like reducing "red-eye,[63] " can cost about $20 for the first person and $7.50 for each additional person in the photo.

Once the retouching is done, it can be printed out at 600 DPI to high-quality laser paper or a regular negative can be produced. From this you can print a photo on high quality photographic paper to the desired size for normal photo enlargement rates.

Digital Imaging to
Move People & Create New Photos

These photos were taken from the one on top of page 93. The women have been repositioned to make attractive individual photos.

[63] "Red-eye" happens in color photographs when a person being photographed looks directly into the lens and the light or flash reflects off the eye back into the camera lens.

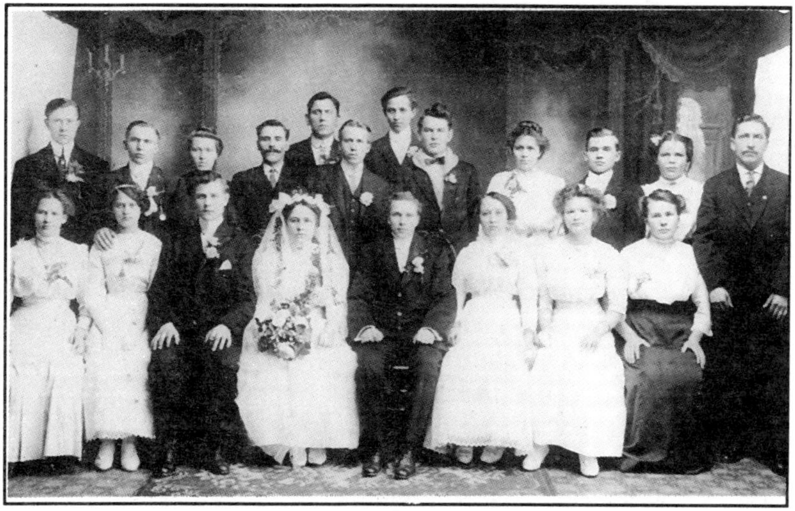

Copy of Original Old Photo

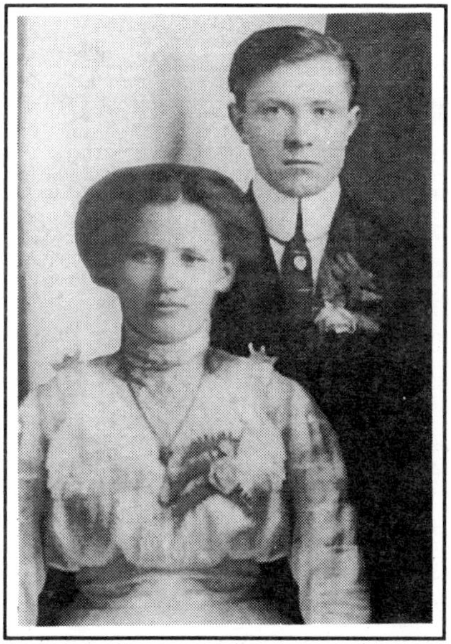

Couple on Far Left

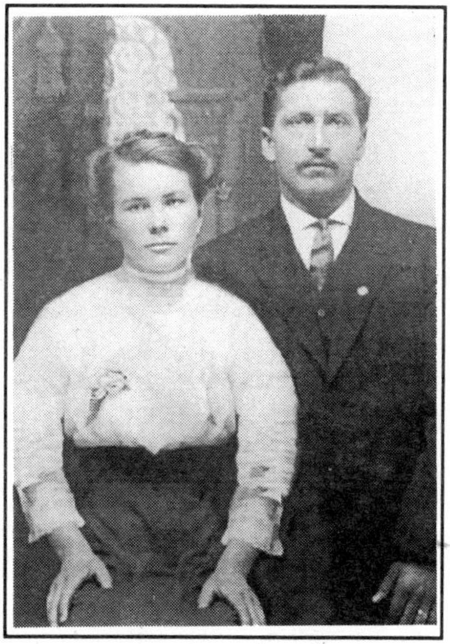

Couple on Far Right

DO-IT-YOURSELF RETOUCHING

You can now use your desktop computer to do many tasks that only manual photo retouchers, printing companies and specialists could do before. You can do anything from the simplest of retouching to creating special effects.

Once you obtain the scanned image on disk, you can do the retouching yourself. If you are a computer enthusiast, you may be interested in learning to do some of these techniques yourself. Results are totally dependant on the software program used and the amount of time you take to learn the techniques.

You must have a good photo software program that can give you the desired results. Be sure the file format of the image on disk is compatible with your software. The most common formats are .TIF & .JPG for photo imaging (.TIF is preferred).

For the first time, people other than artists and specialists can restore or retouch a photo with professional results.

DIGITAL IMAGES PRINTED ON PHOTOGRAPHIC PAPER

Some photo laboratories can take your computer disk (containing the information you scanned to disk) and transfer it to make a photographic negative. From this negative you can make as many prints as needed on regular photo paper.

This is a specialty service that is not commonly available. The cost for this service can start at $12 for each negative, if you can find a company that can do it at all (of course prices will vary).

PHOTO IDENTIFICATION

Take the time to identify your photographic prints. Even if it is obvious to you who is in the photo, it may not be obvious to someone else. It is a good idea to make photocopies (acid-free of course) of your special photos. Put each one on a separate sheet of paper and place the photo at one end making it one-half the size of the page. On the second half, write a complete description of the photo. Include names, dates, places, and the occasion. Put it into story form. Tell your family members (or descendants) why that photo was taken or any interesting piece of meaningful information about the event or why the photo was taken. When you finish all your "special" photos, put the photocopies in a notebook, away from the photos.

There is nothing sadder than inheriting a box of old photos from a deceased relative and not being able to identify them. I have spent years trying to identify photos that were found in my grandmother's genealogy after her death.

A photo that cannot be identified is virtually meaningless. Write on the back side of all photos to identify them. However, be careful with what you use to write. Use a soft lead pencil, never harder than a #2. A film marker is best.[64] Be sure it is not too light to read. (Pencil marks do not show up well on RC paper). Sharp points can crease or indent photos and disfigure the front of the image. Write only on the back edges of the paper, preferably the margin. Never write in the center of the photo in case the ink should migrate and affect the front image.

If you want to use labels on the back of photos be sure they are acid-free or archival, which are not commonly available. Order labels from a company specializing in archival or library supplies. If you use ordinary labels, they will turn yellow and stain the paper (even RC paper) very quickly. The staining of course is because the acid from the adhesive is transferring to the paper. Once the transfer is complete, the label will fall off.

- Don't use a felt-tip maker (Magic Marker) permanent or water soluable on photos or keepsakes.
- Don't use a ballpoint pen or ordinary ink on the back of photos.
- Do use a soft lead pencil.
- Do use acid-free pigment-based ink.
- Do write only on the back edges of the paper, preferably the margin.
- Do use acid-free labels if you use any on your photos.

[64]Dixon Tru/Color Film Marker Black #2225.

LARGE GROUP PHOTOS

A large group is harder to properly identify, but it is very important to do it before everyone forgets who is in the photo.

1. First make a photocopy of the photo, enlarge it to 8x10.
2. Place a blank sheet of paper over the copy.
3. Hold both of them up to a window on a bright day (or use a light table).
4. Roughly trace the outline of each person in the photo. Sometimes it is also helpful to trace the outline of a building or other scenery.
5. Number the people starting at the top-left edge of the paper.
6. Make a list of the names, being sure to use a woman's maiden name and married name.
7. It is a good idea to include approximated birth dates of babies or children, especially if you are not positive about who they are.

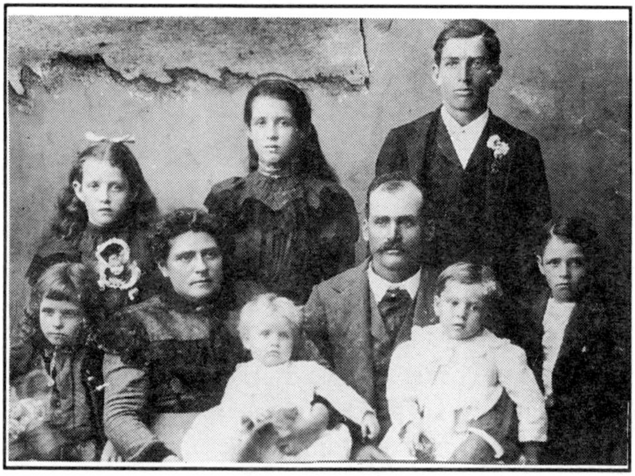

Group Photo
The John W. Smith Family
About 1900
Rocksprings, Edwards County, Texas

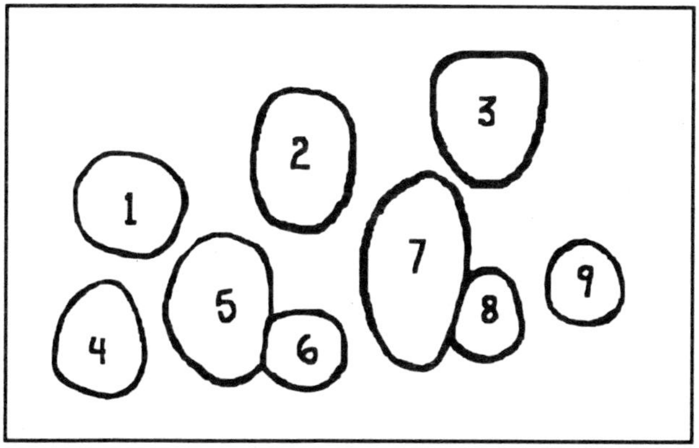

Group Photo Identification

1. Emma. M. Smith, b. abt. Mar 1890
2. Frances Ann Smith, b. abt. May 1888
3. James C. Smith, b. abt May 1884
4. Ida J. Smith, b. abt. May 1894
5. Mary Adaline (Gaines) Smith
6. Eula E. Smith, b. abt May 1899
7. John W. Smith
8. John J. Smith, b. abt. Mar 1897
9. William G. Smith, b. abt. Jan 1892, d. bef. 1910

UNKNOWN PHOTOS

If you are lucky enough to inherit photos of people whose names you do not know, show them to family members and friends before storing them, the sooner the better. Make photocopies of these photos and mail the copies to relatives. They might know who the people are. This could take months before you hear from everyone, but it might be fruitful.

When you are not sure of the names of the people in a photo, it is a good idea to label or identify it that way. Say "This is possibly John Richards, a neighbor, but I am not sure!" You never know when someone may come along and correct or confirm your information.

OTHER WAYS TO IDENTIFY PHOTOS

Close examination of personal and family photographs in an organized and thorough manner may be helpful. You can learn to "read" them with an eye for detail and clues that will help you pinpoint their identity. This is especially true of photos that you know are dated in the nineteenth century (pre-1900).

Our ancestors collected photographs just like we do. They not only had family portraits, but many other photos of friends, local and national heroes, or celebrities. Not every photo will require the same amount of attention. Be careful not to miss that one small but significant detail that might identify a relative or location. Do everything to establish the importance of a photo in an old family album.

Identify the Photographic Process:

❶ Identify materials used for the image, frame or case.

❷ Take careful measurements of image and frame or case.

❸ Note surface qualities, tone, texture.

❹ Use references, books, charts for detail on precise dating. Photography was first introduced in 1839. Tintypes were made from about 1861 to 1930. The size, color,[64] thickness, and quality can pinpoint them even closer.

❺ Write down the description for reference.
—The type of process has much to do with its rarity, value, storage requirements, and kinds of special care needed, if any.

Examine for details, information, and historic content:

❶ Close inspection for details with a magnifying glass if necessary.
—Clothing for time period, uniforms, insignias, fashion, etc.
—Background for tools, architecture styles, signs, landmarks, machines, tools, weapons, banners, signs on buildings, kinds of studio props used, and anything else of interest.
—On mat, case, or frame . . . look for photographer's name, logo, label, stamp, or advertisement. Written information, dates, series numbers?

[64] For instance, brown or chocolate-tinted tintypes were made only after 1870.

❷ Write down findings and consider:
—any further research needed?
—historic, rare, not sure . . . get expert opinion.

❸ Does information agree with the time period of process? Or was it copied from an earlier photograph?

DISPLAYING PHOTOGRAPHS

Photographs are meant to be seen. There are several different ways to display them. You can put them in your wallet, you can frame them and you can place them in albums.

FRAMING PHOTOS

Proper framing sets off the image. Putting a good portrait in a frame is something we all do. We love to adorn our homes with those things we enjoy seeing on a daily basis. There are several things we should keep in mind when we do this.

Anytime we display our valuables you must know there is a security problem. It could be stolen, but you are also exposing it to the dangers of light and the environment.

When framing, a mat board is a necessity in order to keep the print a small distance from the glass. This means when you purchase a frame, buy one large enough for the mat. A 5x7 photo would require an 8x10 frame. An 8x10 photo would require a 10x13 frame, and so forth.

Most photographs and documents put into frames are under glass. The glass protects the photo from dust, fingerprints and other threats. Use glass only, not plexiglass, since some brands could release dangerous substances. All plexiglass is more expensive than glass. Frames that contain glass marked UV glass are best. If, when you put the photo into the frame, you do not do it properly, you could seriously damage the photo or document. Do not trim a photo to fit the frame unless it is a copy. You should never fold a photo to fit into a document. Do not trim a photo to fit the frame unless it is a copy. You should never fold a photo to fit into a frame. Always back your photo with acid-free paper or an acid-free mat board. This separates it from the back of the frame or from frame spacers, which might contain acid.

Wooden frames are beautiful and very common. They are loaded with acid and can damage the photo. Consider using a copy to display. Sometimes you can insert a polyester liner in the frame that will protect the photo. Metal frames are much safer and are used by professional photographers and museums.

DAMAGED PHOTOS IN FRAMES

Photos are frequently damaged after they are framed. A serious mistake that some people make is placing a photo directly against the glass. The photo needs a tiny bit of space to breathe.

If you do not separate it from the glass, moisture will form and eventually the photo will permanently stick to the glass. Moisture collects much the same way it does on the windows of your car when there is a temperature change. Condensation forms on the inside of the glass. If there is not enough room the photo will get wet and eventually warp and stick. Once this happens the surface of the photo is destroyed.

What happens is the moisture develops into mold and the mold affects the photographic image, softens it, and affixes itself to the glass. Identify this damage by what looks like a shinny spot on the photo. If it is a color photo, it may also develop what looks like "bursts" of a gold color (also looks like snowflakes), called "foxing.[65] " This photo and frame were setting on a shelf next to a plant that was watered regularly.

[65] Foxing is caused by humidity. It appears as reddish-brown blotches.

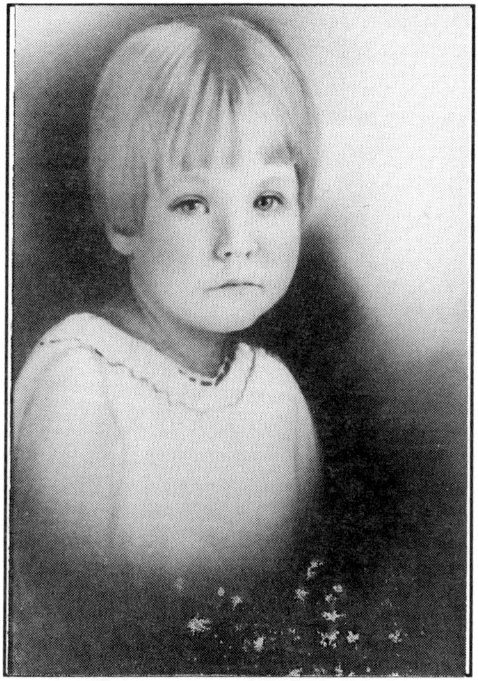

Photo with Mold Damage
Shows snowflake-type (called foxing) effect on a color image. This photo also has many shiny spots that do not show here

Removing a photo that has stuck to glass can sometimes be done without damaging the image, but only if the photo is on 100% paper. It is virtually impossible for you to remove RC paper that is stuck to glass. A professional photographer or curator can do it, but it could be expensive. It is probably less expensive to make a new print. The best thing is to be sure this does not happen in the first place.

One solution to this problem would be to do something to slightly separate the photo and the glass. Sometimes this is done by using an archival mat board between the photo and the glass or some kind of spacer.

MATTING

There are many different qualities of mat board. The most common is the type you can buy at the local department or discount store, photo or art supply dealer. These are the types used for many generations. They are made from heavy paper board, carefully cut out in the center to border your photo. They are available in most colors and are quite beautiful in the frame.

Have you ever taken an old photo out of a frame that had a mat around it? Did you notice that removing the mat often shows a stain on the photo where the mat touched the photo? This stain is where acid from the mat board has transferred to the photo. This happens even when the photo was printed on high quality photo paper.

These stains usually do not ruin the photo. At least it does not obscure the image. But over a longer period of time, this acid can migrate in toward the center of the photo. Using this type of mat should not be done if you want the photo to last a long time. If you are displaying a copy of an old photo, and you are not too concerned with the actual photo, it is probably okay the way it is.

When you purchase mat boards, if they are not labeled as acid-free, archival, conservation quality or museum quality, you should think twice before making the purchase. High quality mat boards are not available at drug, department or discount stores. They are obtainable through archival supply catalogs or art supply stores.

QUALITY MAT BOARDS

Archival mat boards both compliment and protect your framed photos. Mat board made from 100% Rag is the best. If you are on a tight budget, Conservation Board is a good choice. It is both archival and economical.

- **Conservation Quality** is made from purified wood pulp with a low pH. A buffering compound is added. Colors will not fade and are bleed-resistant.

• **Museum board** is specifically made for use with certain types of photos and textiles. It is made from 100% cotton fibers. They are buffered with calcium carbonate, and are not suitable for color photos.

Some colored board will fade with time, but museum board is color fast and will last four times longer. Black mat boards are especially susceptible to dye transfer and should never be used in direct contact with artwork.

If desired, mylar can be attached to a mat board instead of putting it under glass. This method is not totally successful in some cases.

ALBUMS

Essentially albums and scrapbooks are miniature family museums. The first family albums were probably photograph albums that came into existence soon after photography. These albums quickly became the way the world documented, recorded and captured the lives and activities of entire families for decades.

In 1888, when Kodak introduced their first camera, photography became even more popular and the snapshot became the new photographic norm rather than the previous professional portraits.

■ SCRAPBOOKS

The treasured memories of our lives, of family, and friends, such as diplomas, wedding invitations, newspaper clippings, and recognition awards are often placed in scrapbooks. It is not known just how long scrapbooks have been in existence, but probably before photograph albums. The heavy paper used in early albums was very porous and of poor quality.

■ PHOTO ALBUMS

With the snapshot came the familiar black page albums. The pages were black to prevent light from penetrating the pages and fading the photos. These were the most common album in existence for many years. They were inexpensive, and most families owned one. Mass-produced, these albums used cheaper materials, with a high acidic content and harmful glues.

Today we have continued with our own version of the photographic album. Unfortunately, most of these albums are not of archival quality. Many of these later albums used plastic vinyl (PVC) pages that were harmful to photos.

Industry has improved the use of plastics, but the paper and glues are still questionable. Don't expose your photos to the ravages of acid, dust and other pollutants!

■ MAGNETIC ALBUMS

Magnetic albums are the most common albums available on the market. When they were first introduced many years ago, I thought they were the greatest invention that had ever happened. I loved them. They were so easy to use. I bought several dozen of them and mounted hundreds of photos and documents that I wanted to save.

Pages in a magnetic album contain a waxy, sticky backing board. Magnetic album pages look beautiful when they are new. It took me quite a while to learn a hard lesson. After a few years the pages start to turn yellow. You guessed it, the yellow is acid damage. If you look closely at a magnetic album page there are little lines on the backing. What I did not know was this is rubber cement.

After several years, the protective page covering no longer sticks to the backing, especially on pages where photos or paper items are placed. The reason is the acid from the rubber cement has penetrated into the paper items and no longer holds the page cover (film) in place. But if you try to remove the photos, you can't. The rubber cement literally cements them in place. To remove them you have to either tear the photo or the backing. Newspaper clippings are even harder to remove and this nearly always results in their destruction. I call this "double-death" because with the acid in the newsprint and the acid on the page, the clipping does not have a chance.

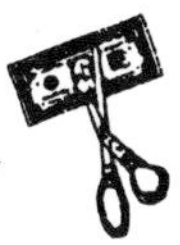

Nearly every family in America has owned a magnetic album. If you own any of these albums, take steps as soon as possible to remove your precious items from them. Don't throw the magnetic album away quite yet however.

If you cannot afford to buy acid-free albums, take some simple steps to protect your photos. Buy some acid-free paper (archival paper would be better) and back all of the photos you have in the album with a piece. Take a glue stick (or acid-free glue) and put just a dot on each corner of the photo and attach it to the acid-free paper. Then you can replace the items in the magnetic album. It will protect your photos from sticking to the page and it will slow the acid migration from the page into the photo itself. This is not an ideal solution, but it will work until you can afford to buy new acid-free albums. Don't buy any more magnetic albums. Warn your friends about this hazard; you will be doing them a great favor.

MOUNTING PHOTOS

The archival method of mounting photos is to use hinges made out of special tissue. Museums use these. It is a lot of work, and most people will probably not want to bother.

To mount photos in albums, it is best not to use adhesives and never use tape. Mounting corners are the second best way to attach photos. Those made of archival-quality materials like polyester film are great, but others will do. Choose the ones that are transparent, so they do not block out part of the photo. Another advantage of these mounting corners is they are repositionable so you can make changes whenever you please.

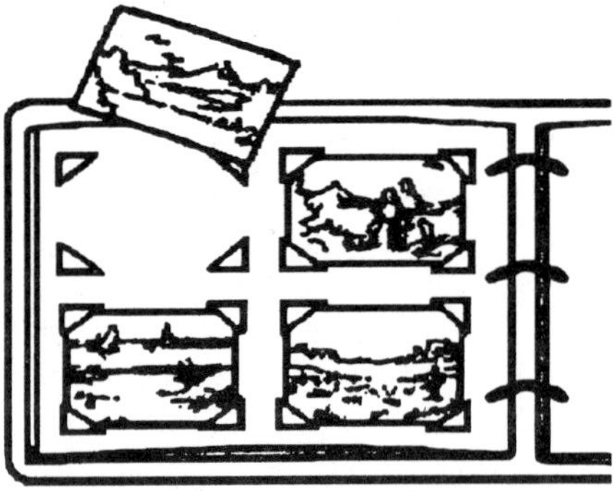

Using Mounting Corners in Album

Another method would be to attach the photos to acid-free paper and cut slits in the paper to insert the corners of the photos. Then attach the acid free paper to the page with an adhesive. The least expensive way to mount snapshots, is to use a dab of glue stick to each corner of the photo and attach it to the page. This is not a perfect way to do it, but if you are careful, it will work with a minimum amount of damage.

Make sure your photos are not "kissing." Photos should always be protected from face-to-face contact. Do this by placing a piece of acid-free tissue or paper between pages of the album. This prevents the images from sticking and transferring. Try not to crowd the photos onto a page. Overlapping them could cause damage to the surface.

In the last couple of years a new industry has appeared on the scene. A franchised business in many areas of the country,[66] they specialize in producing family albums that are archival in quality and creative in arrangement. These businesses are usually operated on a home party-type sales program. The idea is great and the products are also good, but the cost to the consumer is another thing. Items they sell are very expensive. Other than the cost, they have very clever ideas on arranging and putting together an interesting, artistic album that truly reflects your family stories. They teach creative cropping, album page layout ideas, organizing negatives, suggestions for starting your photo album project and much more.

You can do this on your own by learning how to arrange your own pages. You can make the albums more interesting by adding captions that tell the story about the photo.

CLEANING NEGATIVES AND PRINTS

Emulsion cleaner will remove surface oils, fingerprints, some inks, dust and dirt from photos and negatives. Never attempt to clean a photo that is more than ninety-four years old (before 1900), emulsion cleaners could destroy them.

Spray the cleaner on a clean lint free cloth or cotton swab and rub in a circular motion. The solution dries instantly and leaves no residue.[67]

---

[66]One of these companies is *Creative Memories,* another is *Legacy. These are registered trademarks.*

[67]Use emulsion cleaners at your own risk. If used improperly they could destroy or damage prints, negatives and slides.

STORING NEGATIVES

Developing a safe system for storing negatives is vitally important, and not very difficult if you keep in mind a few facts. Negatives naturally contain some acid. For this reason they should not be stored with prints. They should never be stored in buffered boxes or come in contact with buffered paper. Even though they contain natural acids, they should be stored in acid-free containers, envelopes or holders.

Always avoid extremes of temperature and humidity. Protect negatives from dust, direct light, fingerprints, and acids. Glassine should never be used to store film—it is acidic, contains plasticizer and glycerine. Glassine is what old film envelopes and holders were made from. Kraft paper is also acidic. Acid-free papers, Mylar D, and Polyethylene are all recommended for negative storage. Do not allow negatives to come into direct contact with polypropylene.

Envelope design is also a consideration. Seams should be at the edges and outside to avoid pressure points, and adhesives should be pH neutral. Never use buffered paper or boxes for color film.

You should always place negative strips in negative holders to handle. This prevents fingerprints, dust and scratches from ruining the emulsion. They also make these that fit into a 3-ring binder.

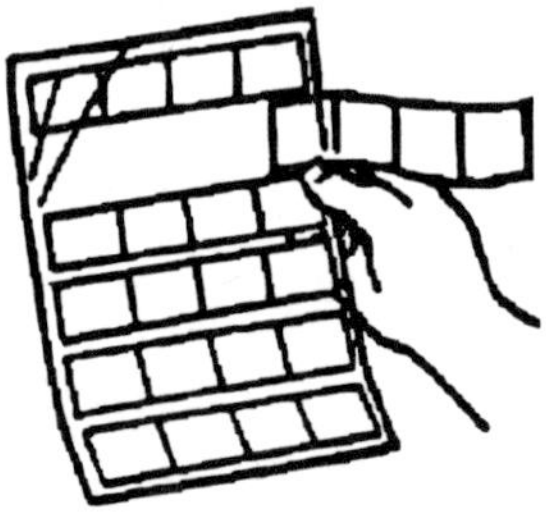

Negative Holder

Negative envelopes should be filed and identified so you can quickly find them when you want to make prints. They should not be stacked together because it will scratch them. Negatives should be separated using acid-free (not buffered) envelopes or polyethylene or mylar sleeves. A plain piece of acid-free paper, cut in half, 8½"w x 5½"h, tri-folded, makes a good economical holder for negative strips.

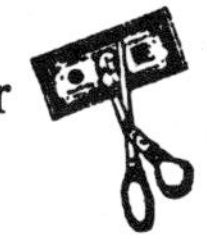

- Don't store black and white film with color film.
- Don't store negatives with prints.
- Don't store negatives in polypropylene.
- Don't store color film or color prints in buffered containers.
- Do store black and white film and negatives in acid-free containers.
- Do store negatives with silica gel.
- Do store negatives and prints in baked enamel boxes or archival boxes.

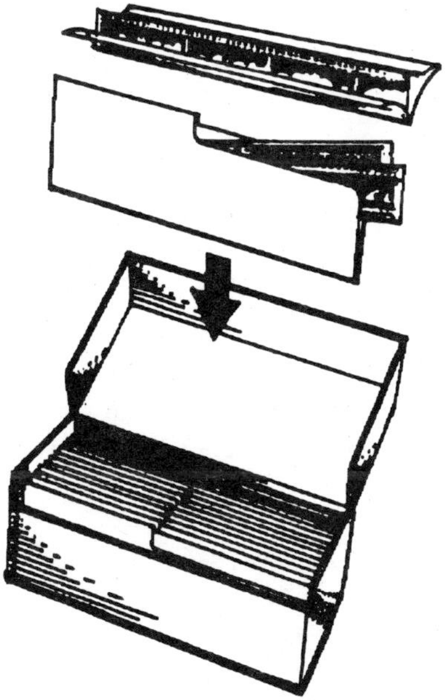

Negative Storage Box

STORAGE OF PRINTS

There are some photos you will want to display, and others you will want to store. Before storing photographic prints, you should be sure they are all properly identified and dated. They should be filed so you can quickly find them when needed.

Be sure you have made your photocopies and written stories about your favorite photos. Try to cover all the information you could possibly know about it. Make it interesting. Someday you will really enjoy rereading it. After you finish with all of the photos, make a second copy and give one to a friend or relative. This could be invaluable to you someday when you want to write a family history, or if you die and your descendants want to know about your photographic collection, or in case of a disaster.

If you live in a hot humid climate, store photos in an inside room or in an air-conditioned room. Store them in a room where you would be comfortable.

Black and white and color prints should never be stored together. They require different conditions to keep them in the best condition. The dye in color prints is more acidic than black and white prints.

Store them in acid-free boxes. Storage in a box of this type guarantees only that the box will not harm the contents. The photographs themselves must be properly processed to avoid deterioration.

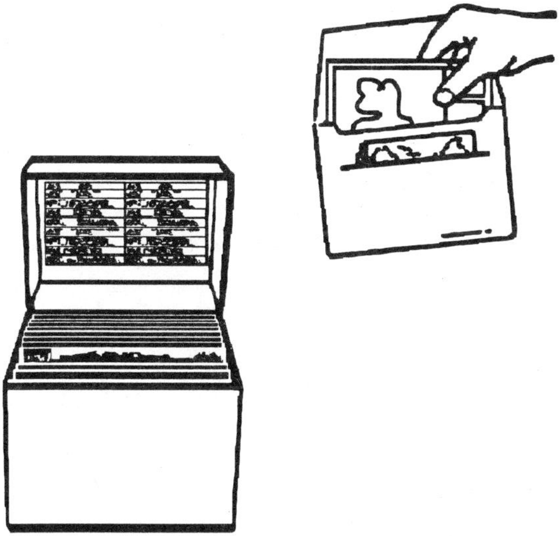

Photo Prints Storage Box

Do not allow the fronts of the pictures to touch each other ("kiss") even when laying flat. When photos are stacked, it is okay for the back to touch the face of a photo, but only if there is no ink on the back. The safest thing to do would be to place a small piece of acid-free tissue (or paper) in between each photo or page of photos.

Many suppliers of archival products produce various types of boxes especially designed to help you store your photos. Arrange prints in chronological order (in or out of albums). They are more interesting that way. If you have photos of other families, arrange them by family group first, then by chronological order. This helps when it comes time to find them.

- Don't store black & white and color prints together.
- Don't store them with fronts "kissing!"
- Don't encapsulate film or prints, especially color photographs.
- Do interleave with sheets of acid-free tissue.

SLIDES

Slide film is more stable than regular film. The colors will remain true longer than color negatives, depending on the way they are stored. They should be stored in baked enamel boxes and the slides should be separated. Archival supply companies make special boxes just for slides. They also sell page protectors for slides. Be sure any that you purchase are polypropylene or polyethylene. If you use these, be sure to check them periodically for signs of moisture.

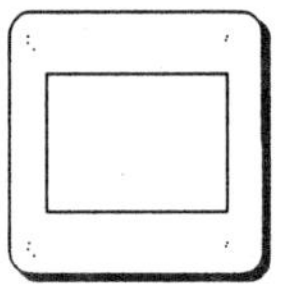

Some people say that the cardboard slide holders can damage slides. Actually, all color film is acidic. The main damage that the slide holders will cause is to themselves. That is, the cardboard will yellow with age. When that happens you could replace them with plastic holders if needed. The main concern with storing slides is you want to protect them from dust and fingerprints. They do best when stored at very low temperatures.

MOVIES AND VIDEO TAPES

Movie film is a color film and is highly unstable and prone to fading and color shifting. If you have old movies, you have probably noticed just how bad they become. Movie film has almost completely gone out of style. Video tapes are the way people do it today. Many people are in the process of transferring all of their home movies to video tape. Duplicating your old movies onto tape is a good idea because it makes it much more convenient to view. It also cuts down on the possibility of damage to the film by improper handling. This automatically prolongs the life of the movie film. The biggest disadvantage is that the reproduction quality is generally only fair on a video tape and they have a projected use life of a limited number of years. No one knows exactly how long video tapes last for certain, but it appears to be no longer than ten years. For this reason all videos should be copied regularly (at least every six years) to be sure they are in good condition, not fading or the tape is not stretching or warping. Keep in mind that each time one video is copied there is a loss in quality (called generation loss). Do not dispose of the original film so you can re-copy it later. **Movie film, even though it fades, is more archival than video tape.**

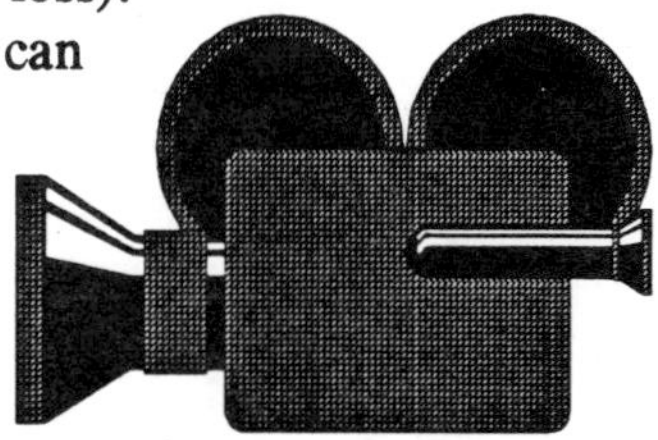

For years we have been told we should rewind tapes periodically. They have now discovered this is not a good idea. It does not help extend their life, it actually wears them out sooner. Every time you play a video tape, more and more snow will appear on the screen. The magnetic particles on the tape actually fall off, leaving a deteriorated image.

Department, television and camera stores sell equipment that will allow you to copy your movies and your slides onto tape yourself. You can also have any photo processing company do it, but it is expensive. I wouldn't be too anxious to rush right out to do this! A new option will soon be available.

COMPACT DISCS

While audio tapes, with care, will last twenty five years or more (archivists now suggest we also do not rewind them) video tapes are much more problematical. One of the major drawbacks of video tapes is their very short shelf life (10 years at the most). This is coupled with the generation loss in making a copy and the damage done even by playing a video tape, it would seem the problems mar the wonderful advantages of video taping. Fortunately, just in time to save us from a serious quandary, technology has been developed that will permit us to convert our video tapes to compact discs, which are virtually archival and will last indefinitely.

Industry uses laser disc technology now to produce compact discs. Although not currently available to the general public, within a short time, these techniques will be here for us too and we will be able to permanently record our images and sounds. Let us remember to convert our movie film and home video tapes to compact discs in the next few years!

OLD FILM

If you have old movies and film, dating back to World War II and before (actually before 1951), you have a serious problem. Old film, made from cellulose and nitrate, is highly unstable and flammable.

About fifteen years ago, I had an old black and white 35mm film that once belonged to my father. It was professional news footage of the time his model liberty ship was launched by a celebrity during World War II. It was a special event that my family delighted in having in our possession. The film was in a film can and I had it stored in a safe, cool, dry storage cabinet. I decided to take the movie out and show it to my children. When I opened the can, I could not believe my eyes. The film had literally disintigrated. All that remained was a mass of gooey cellulose and brown powder.

What I did not know at the time was that this type of film is also subject to **internal combustion**. The film could have burst into flames without any apparent cause. I was lucky to have only lost the film.

If the instability factors aren't a serious enough problem, this older film is also loaded with acid which will not only destroy itself, but will contaminate anything that it touches. If you have any film manufactured between 1890 and 1951 it contains nitrates and is not safe to store.

Professional 35mm motion picture film represents the greatest hazard. Cellulose nitrate film was not manufactured after 1951, but it was still being sold after that date for an additional thirteen years.

It has always been a common practice for photographers to purchase bulk rolls of 35mm film and respool it into smaller rolls for still camera use. This means that not only old movie film must be checked, but also old 35mm negatives.

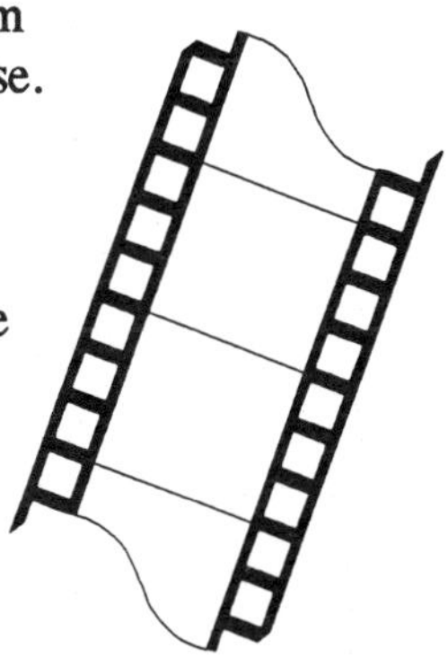

This film also presents a hazard because of the strong acid formed from gases that the negatives release. If you have any negatives or movies that are forty-five years old or more, carefully check to be sure they are on safety film. If not, have copies made immediately and dispose of the original. If you decide not to dispose of it, at least store it in a safe place, away from all other film and flammable items. If the film is marked "safety film" or just *safety*, it is okay.

Once the industry discovered this was a problem, it was corrected by producing a new type of film. Most modern film is now polyester based. It is usually not labeled "safe" as it is taken for granted that we know it is okay. Manufacturers have tested this film and it should be stable for several hundred years, depending on the way it is handled and stored.

Remove your movie film from the original boxes and put them in metal film cans. Each roll should have its own can. Larger metal containers are available for multiple rolls of film. Mark the cans completely so you don't have to guess what film it contains.

- Don't store movie film or negatives that are more than forty-five years old without checking it carefully.
- Don't periodically rewind video (or audio) tapes in hopes of prolonging their life.
- Do store slides in a metal box or polypropylene sleeves.
- Do check stored film periodically for moisture build-up.
- Do store movie film in metal cans, not in original cartons.
- Do copy your movie film onto video tape for convenience in viewing.
- Do transfer all of your movies and video tapes to compact discs as soon as possible.

Proper handling of photographic prints, negatives, slides and movie film, can be a great deal of work. But once done, you will know your family and ancestors with an intimacy not easily acquired with the written word. "Our best photographs can tell us more than the most detailed genealogy about the people who have been members or close friends of our family."[68]

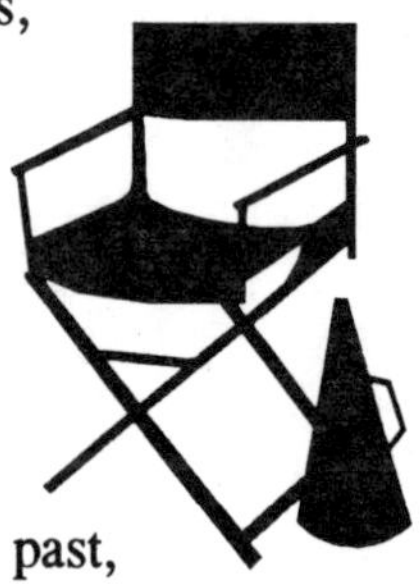

Your collection may be limited to just the recent past, but if you properly identify, organize and preserve your photographs now, your collection will be much more valuable to your great-great-grandchildren.

[68] Thomas L. Davies.

CHAPTER SIX

FABRICS & SPECIAL COLLECTIONS

You may think you have no fabric mementos. Occasionally we do obtain a few items that we might want to pass down to a younger generation. Perhaps it is a wedding gown, handed down from your mother or grandmother. Maybe it is your own gown that you want to save so your daughter can someday wear it. Maybe it is a military uniform that you just want to hang onto for sentimental reasons.

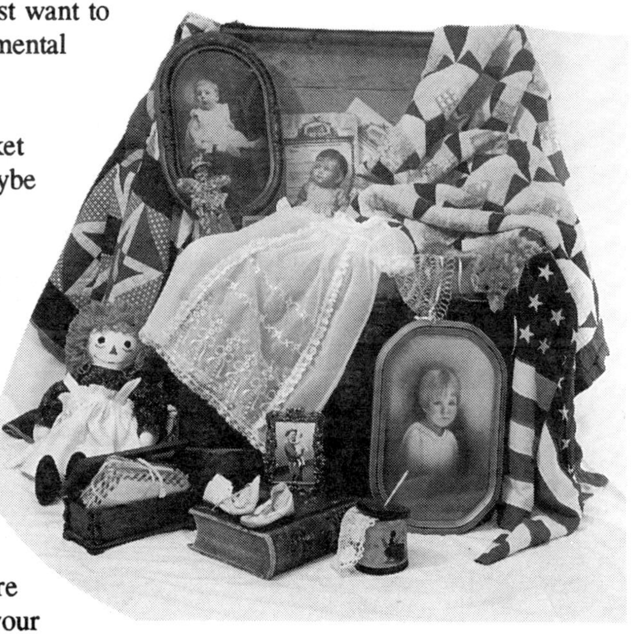

Maybe it is a flag that draped the casket of a loved one. Maybe it is a special linen tablecloth that has been in your family for generations. Maybe it is a hand-made quilt your great-grandmother created. Whether you own an antique lace hanky or a time-worn quilt, here is how to care for your heirloom so you can enjoy it for years to come.

There is a broad range of fabrics that might be involved with your particular heirlooms. These items have certain things in common. They are all prone to wear and tear, they are all sensitive to light, and harsh detergents and chemical cleaners will cause damage.

Preventive care is the first thing to keep in mind, and is the first step to prevent harm to an object before damage occurs. Check them regularly to be sure they are in good condition. Preservation, especially of textiles, is an ongoing thing. It is not a one time procedure.

Check all items carefully, look for small tears, holes, loose buttons or loose threads. Especially check all folds because they are stress points where the fabric could eventually break. Examine the pieces for fading. Improper handling, cleaning, displaying and storage, causes most damage.

REPAIRING FABRIC HEIRLOOMS

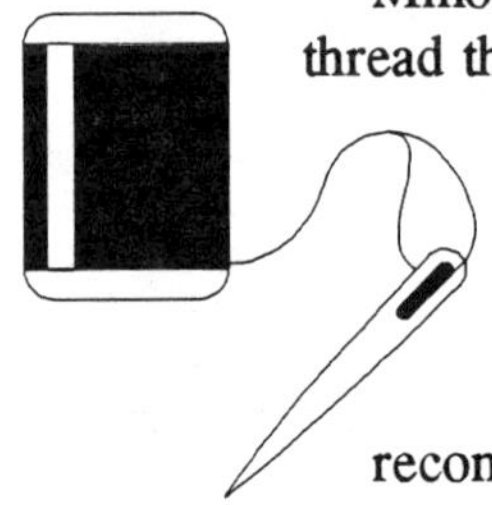

Minor repairs should be made using high quality thread that is the correct color and the same weight as the original. Some people prefer using clear nylon thread. Thread and fabrics fade at different speeds and even if a repair matches now, it may not later.

To preserve the value of an heirloom, experts recommend that you take it to a professional who specializes in restoring fabric antiques. They have the right kinds of thread, fabric, etc. and know how to work with old, delicate fabrics. Restoring damaged fabric is a specialty field, far beyond the average person. Curators in museums receive large salaries just to perform this service. Often it involves re-construction of the garment or item. It can be extremely expensive. Call a local museum or look in the yellow pages for someone that does this work. Get professional advice before attempting any repairs yourself. (See p. 127)

Evaluate your fabric items for their true value. Are they worth the effort and expense of archival preservation? It is far beyond the purpose of this book to go into detail about this. It is better to take precautions to prevent or reduce the possibility of damage occurring in the first place. You can do this by properly storing fabric items.

- Do mend all items before cleaning.
- Do contact a specialist before attempting any repairs to fabric antiques.

CLEANING FABRIC HEIRLOOMS

Many heirlooms over fifty years old may look "dirty" because they have brownish or rust spots. These spots are not dirt. They are the result of a natural action in the fabric and therefore usually cannot be removed. In fact, experts say these spots enhance an article as an antique. So when should you clean your heirloom? Clean it before you store it. Then, clean it only if it is so dirty it cannot possibly be displayed, or if you have recently gotten it dirty yourself. If you decide it is safe to wash, do it very gently with a mild soap (Ivory or Woolite are good), never wring it, and rinse thoroughly.

Satin will turn yellow unevenly. This is more than likely due to the tissue or the box or container with which it has been wrapped or stored. The portions of the satin that come in closest contact with something containing acid will turn yellow first. If your fabric is silk, satin, or wool, it should be professionally cleaned. If it is man-made fiber or cotton, you can probably clean it yourself.

It is nearly impossible to go into detail about removing stains because of the vast variety. Before handling antique fabrics, it is a good idea to wear white gloves, or at least wash your hands thoroughly. Never attempt to take out stains with Clorox or bleaches with "brighteners added." Bleaches of all types are extremely harsh and weaken fibers. They could also shrink the fabric. Some fabrics can use an oxygen bleach (like Snowy), but use it sparingly and test the fabric first. Never wring or scrub the materials. It is best to use a washing machine that does not use agitation to clean but uses a tumbling action (laundromats often have this type washer). Let any antique fabric item dry at room temperature, preferably laying flat on a clean white sheet.

Remember that not all fabric is acid-free. Dyes used to add color to fabric are often acidic. Dyes, especially older ones, were often extremely unstable. If there is any chance that a colored fabric might bleed take precautions before you wash it. To set the color in a quilt proceed as follows: Purchase *Sodium Perborate* at a drug store. Dissolve 1/4 cup to five gallons of water, place it in a new plastic trash can. Soak the quilt in this solution for *up to* three days. Place in washer on gentle, add 1 tablespoon of *Orvus Shampoo.* This is an animal shampoo and can be purchased at a feed store. Wash and rinse for one (gentle) cycle. Place in a large dryer for 10 minutes only. Remove quilt and lay flat on a clean white sheet to finish drying.[69]

Most of the techniques used here for quilts also apply to other items made from fabric.

- Don't ever allow food or drink near the items.
- Don't hang antique fabric or clothing on a hanger to dry.
- Don't ever place it in the direct sunlight or near fluorescent lights to dry.
- Don't ever place it near a source of heat to dry.
- Don't shake antique fabric items when cleaning or drying.
- Do clean all items before storing.
- Do remove all jewelry, pins, or anything that could catch on the fabric and snag it.

CLEANING QUILTS

Is your grandmother's quilt on a shelf in the closet? Quilts were made and saved to be handed down. Avoid using old quilts if you want them to last. They need to be removed from their storage container regularly and aired out. Do not shake them. As quilts age, they may not always react in a predictable manner when we wash them. For this reason, it is a good idea to consult an expert for advice. A conservator will recommend treatments based not only on his or her experience with similar objects, but also upon careful examination of the piece.

The best care for quilts is proper handling and storing.

[69] Information from Beverly Carlson, instructor of clothing design. She also recommends a product named *Retayne*, mfg. by Wild Fiber, distr. by G&K Craft Ind. Box 38, Sumerset, MA 02726.

■ **Airing:** The safest method for a quilt fifty years and older, airing should be done in the shade on a nice, breezy day. Lay the quilt flat on the ground or on a table between two clean sheets. Leave it out for a half day. Never hang a quilt on a clothesline. To air a quilt indoors: hang quilt over a chair or on a bed, leave for a half day.

■ **Vacuum-cleaning.** Materials needed: **low-power**, hand-held vacuum and clean brush attachment, a piece of polyester web, fiberglass or nylon screen with edges taped. (a good size to work with is 24" square.) To do: Lay quilt flat on bed or other clean surface. Place screen on the quilt; slowly and gently vacuum over screen. Move the screen to another part of quilt; repeat vacuuming. When the top of quilt is done, turn back half of the quilt and vacuum the bed or surface under it; repeat vacuuming on other side. Never vacuum a quilt with beads on it. This method should not be done more than once a year.

■ **Dry-cleaning** is best for wool, velvet or silk quilts. It must be done by an expert cleaner who specializes in antique garments. Note: there is no guarantee that dry-cleaning will work or that it will not damage the item.

■ **Wet-Cleaning:** Like dry-cleaning, washing a quilt should be done by an expert. It is too easy to damage if you wash it at home. If the quilt is not old, it can be washed at a Laundromat in an extra large washer in luke-warm water with a mild detergent. Be sure the dye will not bleed (see above for special instructions).

■ **Never press a quilt:** It puts stress on the stitches and could cause serious damage.

The techniques used to clean quilts may also apply to garments such as uniforms, wedding dresses, and so forth. Seriously consider letting an expert do it. Contact your local museum for advice on where to go for help. (See also page 127.)

DISPLAYING QUILTS & GARMENTS

Display quilts for no more than three months at a time. Then, rest them for four months by storing as suggested below. Never hang a quilt or fabric garments on a clothesline or on hangers. This weakens stitches. Damage can occur quickly to a quilt in particular because of the weight of the item. A garment needs better support than a wire hanger or clothes line can give. Only garments in good condition should be hung. A wide hanger with a bottom support to hold some of the weight, may work for a short time. If the hanger is made from wood, it should be sealed before using. It is also a good idea to pad the hanger with several layers of polyester quilt batting to give a fuller shoulder form. Cover the batting with a piece of washed muslin for a smooth finish. A dummy or mannequin (dress-makers form) is good for hanging a garment, at least temporarily. Finally, the hanging garment should be protected with a muslin dust cover made in the shape of a cleaner's bag. Do not use plastic cleaners bags or plastic (vinyl) garment bags which can potentially harm the fabric.

Never allow anyone to sit on an heirloom quilt. It is also a good idea not to put too much weight on them by stacking them or by putting them on the bottom of a stack of other items.

Never display quilts against a wooden wall, or one with wood paneling. If a clean cloth or polypropylene or polyethylene is put under the quilt, it could be placed there for a short time (a few hours).

STORING QUILTS & FABRIC

Storage of fabric, garments, and quilts is a very serious concern if you want them to be in mint condition. Silks, old lace and other textiles need archival protection or they will yellow and deteriorate.

It is best not to hang them on hangers in a closet. Whenever possible, store all fabric items flat. The best way to preserve them is to wrap them in acid & lignin-free (non-buffered) tissue, then put them in an acid free box (textile box). These boxes are quite expensive (depending on the size, could cost in excess of $30 each.)

If an heirloom has a metal part attached to it, remove the metal, if possible, before wrapping and storing. The metal could rust or stain the fabric. Some people remove metal buttons. I think it is easier to take a piece of heavy acid-free paper (card stock), cut a slit in it to the center, and place it under the button. This completely separates the button from the fabric. Replace it whenever a rust stain appears.

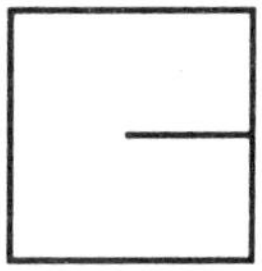

Never store fabric in a plastic (or vinyl) bag or in styrofoam. Plastic can cause rotting if there is any moisture. Avoid storing in anything made of wood, including regular tissue paper. Years ago we were told that blue colored tissue paper was ideal for storing wedding gowns and would help keep the gown white. People did not take long to discover this was not true. The tissue paper not only contained a lot of acid, but the color could actually transfer from the paper to the delicate satin and lace.

Store all pieces in acid-free boxes. Use only acid-free tissue or clean white cloth in the box. Depending on the type of fabric and colors, a buffered box might be best.[70] If you must put them in a dresser drawer, line it with acid-free tissues or washed cotton or muslin sheets. For more protection, coat dresser drawers with polyurethane varnish or other sealer, before lining with the paper or sheets. If your home gets dusty, line drawers with heavy polyethylene sheets, Mylar or fiberfill.

It is best not to fold fabric items at all, but some are too large not to. Use as few folds as possible. If it is not possible to fold, roll it gently, and move it often. The best way to fold a quilt for storage appears below. This method prevents soil buildup, discoloration and wear along fold lines.

Fold the quilt into thirds, placing a roll of acid-free paper along and under each fold. Now fold the quilt again into thirds, toward center. Place a roll of crumpled acid-free tissue along and under each of these fold lines. You can substitute washed sheets, unbleached muslin, white

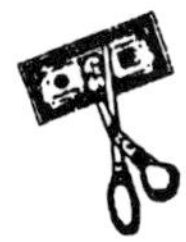

[70]Do not use a buffered box for colored fabric. The buffering compound damages dyes.

pillowcases or terry cloth towels. Muslin is perfect to store fabric keepsakes.[71] It is an inexpensive substitute for the acid-free paper.

Many dry-cleaning establishments offer a special service of preparing a wedding gown and similar items for storage. First they clean the garment and wrap it in tissue and put it in a storage box. This service ranges in price from $50 and goes up. It is usually not archival storage however. They use a beautiful box, sometimes a double box, the inner one having a see-thru top. The boxes and tissue could be dangerous. Be careful and know what you pay for. True archival preparation for a wedding gown can cost $150 and more.

Rolling fabric items for storage is also an option, especially for larger pieces, such as quilts and rugs. Archival tubes are available to give the best support for rolled items. A plain cardboard tube can be used if it is buffered by several layers of acid-free tissue.

- Don't ever allow wood to come in direct contact with fabric.
- Don't use ordinary tissue paper, it is highly acidic.
- Don't ever store it in plastic.
- Do store with acid-free tissue or clean white cotton cloth.
- Do use an acid-free box.
- Do use special precautions if you cannot afford archival storage.
- Do try to keep it as flat as possible, with as few folds as possible.
- Do store it in an inside area of the home, not in the attic or garage or against an outside wall.
- Do protect all fabrics from mice and insects. Mice love to use fabric for their nests and you know what moths do.
- Do reposition fabric items occasionally to ease tension on fibers.

[71] Make sure the white cloth is clean—wash it first in a mild detergent and rinse thoroughly.

LACE

For a very mild cleaning of lace or linen, soak each piece in distilled water at room temperature for a half-hour. This removes foreign matter and loosens dirt. Before washing lace, trace it on a piece of paper. If you are washing an all-lace item, lower items into water and take them out on a fiberglass screen with taped edges. For extremely fragile or very lightly soiled items, dissolve 1/8 of a 3.2 ounce bar of Neutrogena or a small amount of a mild laundry detergent in one cup of distilled water at room temperature. Add one ounce of this solution to a gallon of distilled water. Soak item for half an hour, making sure to rinse thoroughly afterward with distilled water. If it needs to be bleached, use lemon juice in water or diluted hydrogen peroxide. Don't leave it in the bleach very long. Rinse well.

Dry the lace on the piece of paper using stainless steel pins to reblock it to its original shape. Place the paper on undyed terry cloth or plain cloth towel and let it dry at room temperature. If it shrinks, do not force it to stretch, you may tear it. Do any ironing by hand. Steam-press lace on a well-padded ironing board, placing a damp undyed (white) cloth over lace first. For linen, start with iron set at cool. Continue raising heat and testing fabric on wrong side until you reach the setting needed. Iron item on wrong side, placing a clean, damp undyed cloth between it and the iron. To iron embroidered linen, place it on a doubled white bath towel to cushion the threads so they will not tear. Store lace flat, in acid-free tissue and an acid-free box. If it has to be folded, insert safe tissue in the fold to soften the edges.

- Do dry by spreading on traced outline and let dry at room temperature.
- Do any ironing by hand at the coolest possible setting.

NEEDLEWORK

Be careful not to wash or clean needlework because the threads can be weakened and unravel or deteriorate.[72] If it gets dirty, have it dry-cleaned by a reputable company. Needlework and lace are both affected by pressure caused by weight being applied. To protect needlework, cover with glass. Leave space between covering and item so threads are not crushed. If you frame needlework, lace or linen, always place it on acid-free mat board between it and frame backing. (See also page 134).

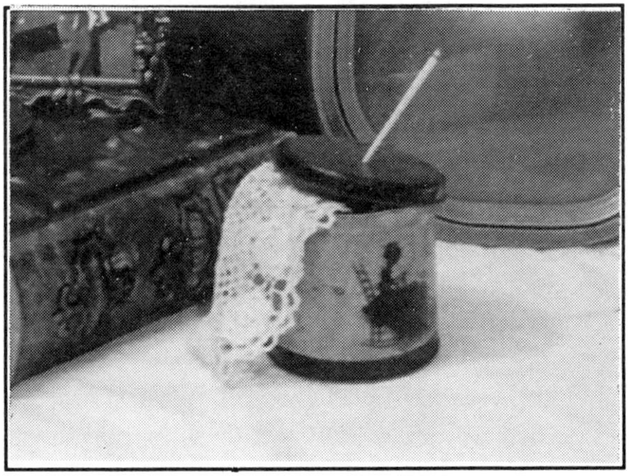

- Don't display an heirloom in direct sunlight or its reflected glare, near heat or under fluorescent lights.
- Don't place any object on top of an heirloom, even one under glass.
- Don't wash or clean needlework.
- Do place on acid-free mat board when placed in a frame.

[72] Some of this information was taken from "How to Preserve your Heirlooms" written by Patricia Mlenak.

ARCHIVAL SUPPLIES

Use some caution and do lots of label reading when purchasing materials for display and storage of keepsakes. Just because you think a product is safe, it may not be unless it is clearly identified as archival, acid-free, or neutral pH. Purchase the best archival quality boxes for long term storage. Select unbuffered boxes for most fabrics, uniforms, costumes, and quilts or any fabric with color. If you cannot afford to do this now, store item wrapped in acid-free tissue or white cotton sheets or pillowcases, or better yet, unbleached muslin.

Art supply stores, department stores, and even camera or photography supply stores, may or may not stock safe items. For that matter, archival supply companies often sell many unsafe products. If it is not marked, it is probably unsafe for long term display or storage. Read the catalogs or labels carefully before you make a purchase. Buyer beware! Some individuals will say anything to sell their merchandise, they have even been known to mislabel products.

Archival supplies are relatively expensive, but a worthwhile investment for your collection.

- Do purchase the best archival quality boxes for long term storage.

WHERE TO GO FOR HELP

For computer enthusiasts, there is a forum on the **Internet** about Preservation and Conservation. It has a question and answer area with input from the "professionals." Keeps you up to date on the latest advances and techniques for preservation.

To locate a conservator for advice and to treat fabrics, contact the Conservation Referral Service of the Foundation of the American Institute for Conservation of Historic and Artistic Works. It is a professional organization to which conservators belong, and their free referral service will give you a list of individuals in your area who are qualified to undertake conservation treatments. A brochure with guidelines on how to select a conservator is also available.

Conservation Referral Service
Foundation of the AIC
1400 16th Street, NW - Suite 340
Washington, DC 20036

SPECIAL COLLECTIONS

Special collections include items like *ephemera* which is a term used to describe things such as coin, stamp, post card and baseball card collections. All of these collectibles need special storage containers. Because these have a specific value (depending on the market), very special care must be taken to keep these in the best possible condition. Consult specialty stores and collectors. Basic environmental conditions are of great concern as with other keepsakes. If you are using less expensive older products for containers, envelopes, and page protectors, you need to change over to newer, safer material.

- Do invest in appropriate archival quality page protectors and boxes designed for the specific type of collectible.

CHINA & GLASSWARE

Never wrap china or glassware in newspaper. The ink may transfer and permanently stain the china. For long term storage, wrap it in acid-free tissue. Do not wash it in the dishwasher because the harsh detergents will destroy decorative gold edges and delicate hand painting. If a piece is broken, do not glue it. A professional can re-fire it in a kiln to fuse the pieces back together, but not if there is glue present. Avoid repeated washing or scrubbing of original painted surfaces such as those on hand painted china, glass or tin.

Allow space in between pieces of glassware. Glass expands and contracts, depending on the temperature and needs room to breathe. Excess moisture could cause the glass to become cloudy. Never store glasses with the bowl side down. This causes moisture to build-up and could damage the glass. Fine crystal can be cleaned with vinegar and water.

FURNITURE

The cardinal rule in preservation is not to permanently change an item you want to have last. It therefore should not surprise you that this rules out refinishing furniture or refurbishing objects in the hopes of enhancing them. Modern repairs often destroy old finishes on

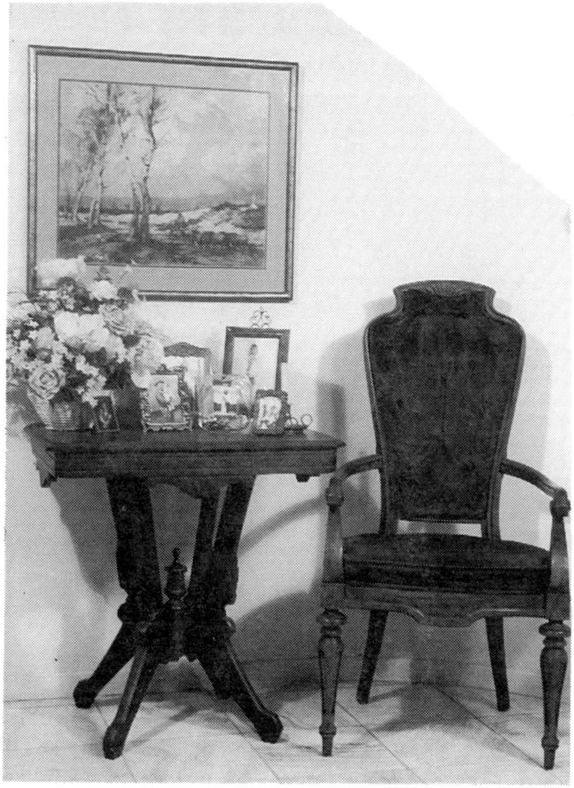

antiques. Preserve furniture handed down in your families in its original or near original condition. Do not even replace or change knobs on a dresser or chest of drawers. It could affect the value of the piece. If you are uncertain whether or not you should repair or refinish an heirloom, consult a specialist.

ARTIFACTS

Store most artifacts and mementos by wrapping them in acid-free tissue and storing them in acid-free boxes. In nearly all cases, it is best to keep the original box or carton in which the item came. These are not acid-free, so you must take precautions to keep the carton from contaminating the item. The original boxes have value and will increase the value of your keepsake. Make as few repairs to your mementos as possible. Repairs to items like dolls, should be limited because it could cut down on their value. Keep in mind, you want it to be in as near original condition as possible. Keep them by protecting them from air pollutants, moisture and temperature fluctuations.

If you have a collection of miscellaneous artifacts and want to store them in special acid-free containers, companies make some that allow you to separate items into box trays that are divided into four, six, nine, or fifteen sections and then the trays fit inside regular storage boxes. Each tray has metal reinforced corners.

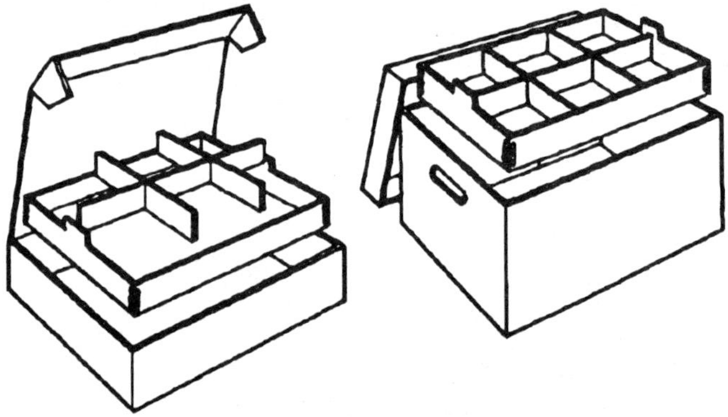

Artifact Storage Boxes

Other acid-free boxes are available for most odd shaped items. You can make a box yourself but be sure it is made from safe material.

Display many small items, like war medals and thimbles, in shadow boxes under glass. Take the same precautions as you would for fabrics. Because of the metal, be sure these are laid flat so the ribbon will not get stained.

The best way to store artifacts includes wrapping them with acid-free tissue, then storing them in any safe container. It is usually a good idea to keep individual items separated so they don't scratch or contaminate each other. Be sure to write up the history of the item, include a list of the special events surrounding it and why it is important to the family or as a collection.

CHAPTER SEVEN

GENERATING AND PERPETUATING KEEPSAKES

You say you don't have any keepsakes? Maybe you should create some. Every family has keepsakes of some type. Maybe you just need to take a look at your collections to decide what really is important and what isn't.

If your home and possessions were in danger of destruction, what would you most like to save (not including family members, pets, cash, keys, wallet, etc.). Make a list of your choices and arrange them in order of priority.

Family Bibles, personal journals
Birth, marriage & death certificates
Family photos, albums, snapshots, scrap books
Family histories published & unpublished
Genealogy records, research notes, source documentation
Wedding gowns, christening dresses, uniforms, costumes, flags
Heirloom linens, quilts
Crafts or handiwork, needlework
Your favorite pieces of jewelry, even if they have no great dollar value
Collections of coins, stamps, baseball cards, etc.
Computer disks, video tapes, audio tapes etc.

CREATING YOUR OWN KEEPSAKES

Newly added heirlooms are just as important as those of your ancestors. Ask yourself "What is important to you that will fascinate the future?"

PERSONAL JOURNALS AND HISTORIES

Write a journal. Preserve your personal thoughts and ideas on a daily or weekly basis. Try to be consistent about it so it has some continuity. Write your personal history. There have been many good books published on this subject. LDS bookstores have a good assortment of them at all times. Once you do your own, then write personal biographies of other family members. Someday these could be your most valuable personal possession.

FAMILY HISTORIES

Do your family genealogy. If you have not already started, do it now. Involve other family members. Do not procrastinate. Living relatives can give first-hand, eye witness accounts which may never be found elsewhere. A life time of "tomorrows" may be needed to find the answer to a question that you should have asked "today."

Publish the results of your genealogy research in book form. It doesn't have to be a large book. Even a small booklet could become a real treasure over the years. Preserving your family heritage is more than preserving documents, books, photos and miscellaneous keepsakes. It is also writing and publishing your family history. Computers make it easier today than it has ever been.

There is a big difference between a family history and a genealogy. A family history is a story about a family, with as many biographies and personal information as could be collected. A genealogy is a record of the names, dates and places of members of a family with little personal detail. A family history focuses on the lives of ancestors rather than on names and dates alone. It adds dimension to filling in between the birth and death dates' of our ancestors lives.

FAMILY ASSOCIATIONS

Create a family association, hold regular reunions, and create memory books of shared activities.[73] With all the time and energy that goes into planning a reunion (not only family reunions, but also school and military reunions, etc.), you may want it to be an event that folks remember for a long time. A jacket, t-shirt, patch, hat, pin, or mug that contains a reunion logo may be appropriate to keep as a keepsake. Take lots of photos and videos at these events that can help everyone relive the event in years to come.

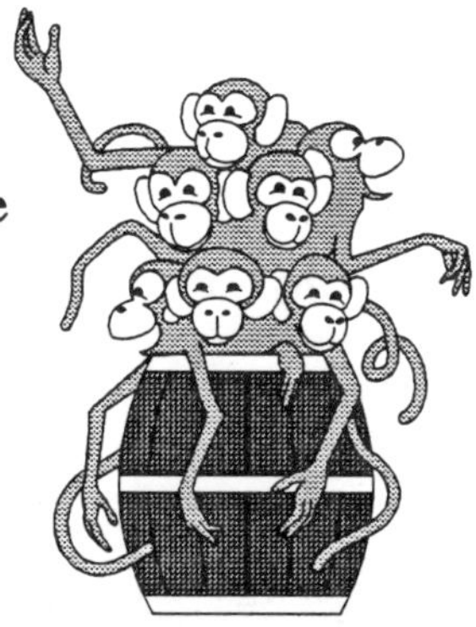

Family Reunions are a Barrel of Fun

Family association newsletters are a wonderful way to share genealogy information and keep up to date on family news. Make sure someone takes the time to do an index at least once a year. Save extra copies of the newsletter. After a few years, have them bound (including the index) and give a copy to a genealogy library in the area where your family home is located. Keep a copy for your own collection.

MEDICAL HISTORY

Recording the medical history of your family could be important information for our descendants. What a great idea to make a "health inventory" for at least your own family. Keep a copy with your important documents and one with your genealogy records for future generations of the family. Include information about your aunts, uncles, cousins, as well as direct lineage and children. This information may take quite a while to collect but could be extremely valuable someday. Include as much detailed information as you can

[73]There have been several good books written about how to form a family association and how to organize successful family reunions and should be consulted for detailed information.

collect. Sample entries to show the range of information that could be included might be:

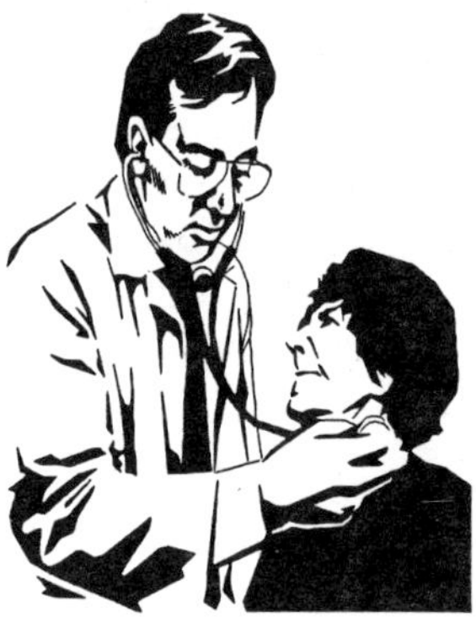

John Lee Miller: paternal uncle, living at age 95. Alert, active and living alone until age 90 when he had an apparent stroke. Now (Jan 1995) living in convalescent home. Moderate diabetes controlled by diet, diagnosed at age 55. Cataracts both eyes age 87, right lens implant at age 90, successful. Several small strokes with residual left side weakness in 1993.

Susan Chandler: sister, living at age 47. Left-handed, red hair and blue eyes. 5'11" tall, average weight. Good health. Had all childhood diseases as youngster with no complications.

Marnell Miller: adopted daughter, came to our family at age five days. She was born drug-dependant and suffered episodes of apnea and bradycardia for twelve months. She needed a monitor for the first year and a half. Brown eyes, she was bald first two years, then blond (almost white) natural curly hair. Broken left leg at age four. Chicken pox at age five, with one small scar on her face above left eye. Dog bite on face with three stitches at age six. All childhood immunizations at prescribed ages. Tested positive for TB in first grade. Took medication for several months. XRay showed no active disease (exposed during kindergarten).

NEEDLEWORK

Other ways to create an heirloom could include: Preserving your precious photographs in cross-stitch, petit-point, or needlepoint. In addition to photographs you could reproduce a likeness of the family home, a favorite poem or saying, or family pedigree chart. (See also page 126).

VIDEO AND AUDIO TAPES

Copy all of the old family home movies and slides onto video tape and make copies for the family or better yet, record them onto compact discs (see also pages 112-113). Create new videos at family gatherings and special occasions. Tape (audio or video) a family member talking about family keepsakes, telling about their origin, history and why it is in our collection.

You should also make videos and audio tapes to record a person recounting his life or special occasions. Oral histories are extremely important as a way to preserve our heritage. Pretend it is an interview for television and ask interesting questions. There are good books available on preparing oral histories.[74]

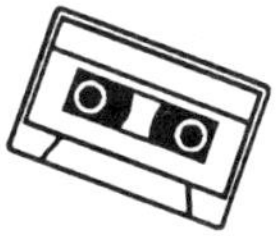

WHAT ELSE ?

Anything having sentimental value to you can become your own keepsake to hand down to your descendants. Thus, you can designate your own belongings as future heirlooms if you so choose. Pass your treasures to the next generation well-documented and in good condition. Link your family history with objects you consider keepsakes. Teach your children to appreciate heirlooms so they remain in the family.

While delving into your family's past, you might consider leaving more then a few traces for future generations.

Be a good ancestor. Do all those things you wish your ancestors had done! Items don't have to be expensive to be worth designating them as heirlooms.

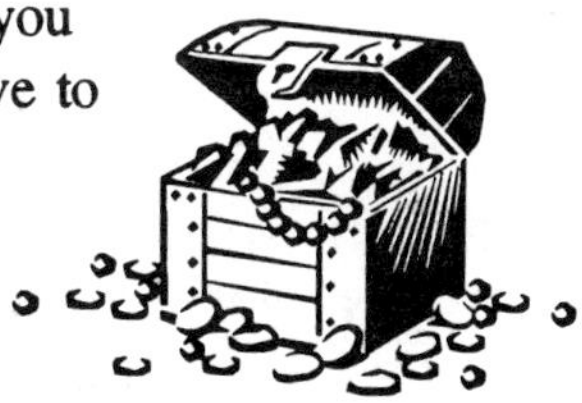

[74]Gary L. Shumway and William G. Hartley, *An Oral History Primer*. (Salt Lake City, UT: Primer Publications; 1973.

PERPETUATING KEEPSAKES

Owning keepsakes and heirlooms gives you a peek at the personalities of your ancestors. Many families possess a few things handed down through the generations. Others have inherited nothing, but they have acquired possessions they sentimentally value and want handed down.

Curiosity about an heirloom can prompt interest in family history, making genealogists out of people who inherit heirlooms. Regardless, a written record (or audio taped story) about an heirloom contributes to family history even if the present owners make no claim to being genealogists.

Genealogists should make a point to visit antique shops in the areas where they research or where the family once lived. Ask local antique dealers and auctioneers to watch out for items having your family name. You never know when you may be able to find an antique that once belonged to your earlier ancestors.

SELECTING TOMORROWS HEIRLOOMS

Maybe your family has no idea how you feel about certain possessions in your home or that you are concerned about having them available to hand down to descendants. What will happen to your treasures when you die? Having no children or children indifferent to genealogy doesn't mean no one cares. There may be someone you never thought of, a distant cousin, niece or nephew, or maybe a yet-unborn great niece.

REFRESH MEMORIES

Make sure family members know and care about the items by refreshing their memories. Periodically remind them by talking about them or telling stories about them. Young children especially enjoy hearing stories about your treasures. Go through the keepsakes one item at a time with a new daughter-in-law. Talk about the items, it is a wonderful opportunity to help someone get to know you.

DISPLAY YOUR HEIRLOOMS

Create a special place in your home to display the items. Usually this is a curio cabinet, but a shelf, a special room, etc. work well. Once you decide to display your keepsakes, it can be said that a buried treasure is now accessible, not only your family and visitors to your home, but also to the next generation. It is nice to enjoy your collectibles. See previous chapters for help on how to display with safety.

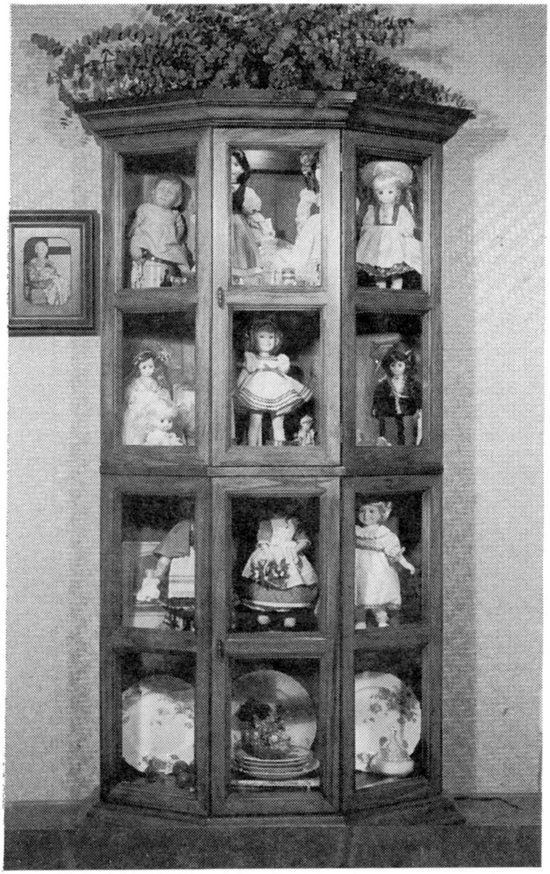

Display your collectibles!

SHARE

Share your treasures by making copies[75] of valuable documents, genealogies, photos,[76] computer disks, video tapes and audio tapes. Give the copies to relatives who do not live with you. This could be important in case of a fire, flood or other disaster. These could also make very nice gifts.

KEEP INVENTORY LIST

It is important to make a complete inventory of all items in your home that you consider to be keepsakes. List what they are, where they are, and information on why the item has a family significance. It is also good to include a date of ownership, how you came to have it, who gave it to you and their relationship to you. Be specific, don't just say, "my grandmother"— you have two of them! Give copies of this inventory list to your children and your siblings.

In addition to the list, write a short history of each item and keep it with your important papers. If the item is large, tape (with archival tape, of course) the history to the underside or backside so it stays with the item at all times. This history should include as many facts as you know about the item, but the most important thing is to tell a story on how it relates to you or to the family. Tell why it is important or cherished. This information could also be included, along with a photo of the item, in a published family history or in your personal journal.

This project could take quite a while. Do not try to do it all at once. Just get started and do it. If you just work on it for an hour a week, you would eventually have a wonderful list and history of all your keepsakes in a short time. This is a wonderful project for someone that is confined to the house, or to a bed because of an illness or accident.

[75] Be sure to print them on acid-free paper and use a copier that has a powdered toner.

[76] The best photocopies of photos can be made on a color-laser copier. They almost look like the original.

LABEL OR TAG ALL ITEMS

Attach an acid-free label or tag to identify all significant keepsakes. Each label or tag (the kind attached by a string) should include complete information about the item. If the item is too small for this kind of tag, use a label and attach it to the bottom or inside. It should include even more detail than the information on the inventory list. Every heirloom has a story, and the items become more valuable if the story is attached. Unless it is written down, someone will forget, and the item becomes insignificant and unappreciated. Record special notes and tell about special family traditions connected to the item.

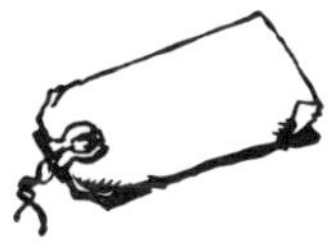

If you cannot properly identify specific items, explore antique shops or libraries for books on antiques and look for similar items.

PROVIDE REASONABLE SECURITY

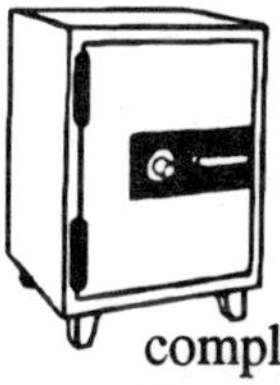

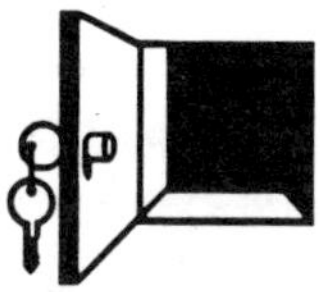

Just because an item is old, does not mean it has a great dollar value. Don't display originals, it is too easy for them to be stolen (or borrowed!). Consider using a safe deposit box for the most valuable items, especially expensive jewelry, bonds, etc. Fireproof boxes are great, but they do not guarantee complete protection. Last year in the terrible fire-storms that hit Southern California, a friend lost everything including the documents she had carefully stored in a "fire-proof" box. The box was intact, but the seals and handle were destroyed as were its contents. Plastic containers will melt in a fire. Consider investing in a good quality fireproof safe, and when storing items inside, keep them away from the sides which would get the hottest in case of a fire..

INSURANCE

Insure your heirlooms where appropriate. Sometimes, however, the cost of insurance makes you think twice before listing all the items you consider important. Some of your treasures have no dollar value because no amount of money could replace them. Some collections have a large dollar value. Documents, certificates, and photos are usually not insurable—nor would you want to pay for insurance.

Not all insurance covers things like floods and earthquakes. If you are interested in having it, you must request it and it can be expensive. Your home owners policy does not automatically cover all of your keepsakes. Any special collection or heirloom that has a specific dollar value must be added onto your policy.

When you purchase insurance, ask lots of questions. Make sure you know what is and what is not covered. You don't want any surprises in case of a loss.

EASY ACCESS

They should be altogether in a spot that is easy to get to in case of fire or disaster. If you were told you had only five minutes to remove all your valuables, would you be able to do it? Rank them is order of importance and let everyone in the household know which items should be saved first. It would not be a bad idea to make a list and post it someplace (inside a closet door, etc.).

DISASTERS

We are constantly being warned to be prepared for the great earthquake. We all live with the possibility of many other disasters, like fire and flood. The best protection for your belongings at such a time is preparation. Many good pamphlets and books are available on how to be best prepared for these impending disasters. If you follow the advice in this book you will make copies of your most valuable documents and photographs, you will make extensive inventory lists of photos and keepsakes, heirlooms and collectibles. You will publish the results of your genealogical research. You will make copies of your computer disks, audio tapes and video tapes. Most of all you will give copies of these things to friends and relatives out of the area.

LET IT BE KNOWN!

Let your family know what you want done with your keepsakes and treasures in the event of your death. Make your feelings well known so no one can do otherwise after you are gone. Make an audio or video tape of yourself telling about your collectibles and what you would like to have happen to them and to whom you would like them to go.

INCLUDE THEM IN YOUR WILL

Include your most valuable treasures in your will. Be specific and include a copy of your inventory list.

I have a friend, a little older than myself, who has put labels on the bottoms of all her possessions. The label has the name of the person she wants to have it when she dies. Of course if your family is prone to squabbling, it may cause hard feelings, if someone disagrees with what is to be done. Perhaps it is better to handle this now, while you can explain your feelings to your loved ones.

BE FAIR

In an attempt to be "fair," when it comes time to divide memorabilia that belonged to a loved one, frequently items do not go where they should and valuable keepsakes are lost.

"A similar problem occurs when a father or mother dies and the diaries he or she kept over the years are divided among the children.
Dividing up a set of journals is every bit as destructive as taking a single journal and ripping it apart so relatives can each have some pages. Like broken Humpty-Dumpty, separated diaries rarely get back together again.[77] "

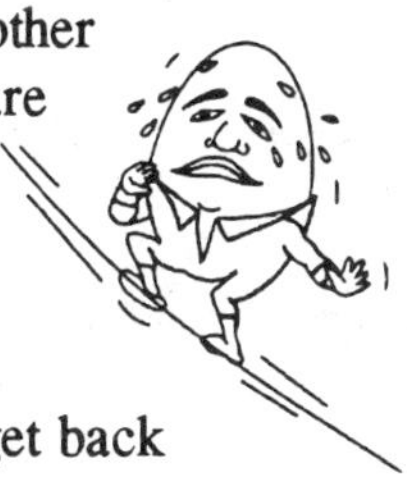

[77]William G. Hartley, *When in Doubt, Don't Throw It Out.* Ensign, 1986, p. 52-55.

WHAT IF NO ONE WANTS THEM?

What do you do if no one in your family is likely to want your keepsakes? Save them from the trash or fireplace. All too often, after a person dies, a well-meaning relative is delegated the responsibility of boxing up and disposing of personal possessions. I have seen it happen all too often. It is usually not until after the fact that others find out that it was tossed into the trash.

Maybe a more distant relative would want them. Get to know your relatives, cousins, nieces, nephews, and more distant relatives. Wouldn't you like someone to get them that would truly appreciate or use them?

If the items are of museum quality or have a historical significance, find a historical society or museum in your area that might want them. Don't be disheartened, however, if someday they decide they don't have room for them and they decide to sell or dispose of them.

Maybe you would consider selling them. Unless you have a valuable estate, it is likely they might be sold at a garage-type sale at a mere fraction of their value.

If it is a collection, (coins, stamps, etc.) consider contacting a collector or dealer to find out its value. Sell to the highest bidder or take the best offer. If you decide to do this but don't want to part with it until you are gone, state in your will that you want it sold and include instructions on how to go about selling it.

Genealogical items are often discarded by family members who never appreciated or understood your interest in the family tree. . . Many times these records are not well organized and look like it would take a lot of work to get them into any form someone else could use.

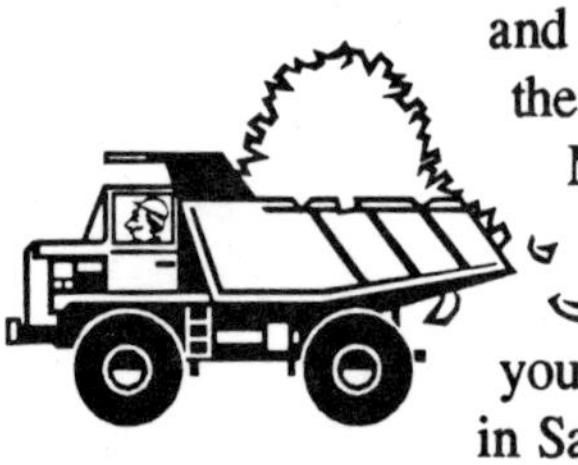

Many times a genealogical society might be interested. You could contact the one in your area and ask. Maybe they could refer you to someone else. The Family History Center in Salt Lake City, Utah, may be of some help here.

If you are just starting to make your history for tomorrow, you should begin to organize your collections today! Don't wait until you have several years worth of keepsakes piled up. Start now. One of the biggest stumbling blocks for organization efforts is the prospect of categorizing the numerous documents, photos and heirlooms from past years.

If you have decided now, after twenty five years, that you're going to get organized, my heart goes out to you, but persevere. It will be worth the effort. Take it slowly and carefully. It is not an impossible task, just time consuming. A neatly organized collection is much more likely to survive than a jumbled mess.

Being a genealogist and family history enthusiast myself, this is really an important topic to me. For that reason I have done everything I can to make sure my collection (if you want to call dozens of file cabinets, tons of paper and a life's work, a collection) will be dealt with properly, and I am not leaving that responsibility to my family, even though they repeatedly assure me they know how I feel. I have entered the majority of the information into my computer (Personal Ancestral File, P.A.F.) and I have published several family history books on the various lines I have spent so many years researching. I plan to dispose of all but the most important pieces of paper.

In general, keep heirlooms in as near perfect and original condition as you can. If a keepsake stays in a family, it can please, remind, even fascinate generations to come.

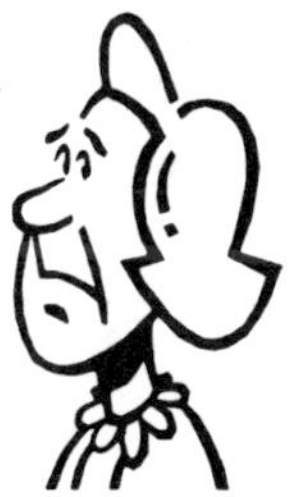

Don't let all this information overwhelm you. Now that you know the facts, you are prepared to do a good job of preserving your family keepsakes. You can make informed decisions about what you can realistically do with your keepsakes. Not all of us can afford to purchase the many archival products it would take to properly handle our items. Take intermediary steps. Do what you can, tackle a little bit at a time, with archival treatment as your goal. Do it as a family project, on family home evenings, and reminisce about the past. Have fun doing it.

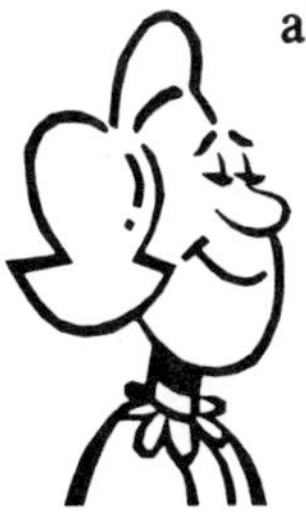

CONCLUSION

CONSIDER YOUR OPTIONS

Archival preservation is ideal and the best possible solution but it is expensive. What do you do if you cannot afford true archival treatment? You could ignore the advice in this book. By ignoring what I've said, you can take the avenue of least resistance and keep doing whatever it is you have been doing. Maybe you will die before any serious deterioration takes place. Or you could take a few easy steps to at least minimize damage. You could make informed decisions by considering your options, using your common sense and benefitting from the experts.

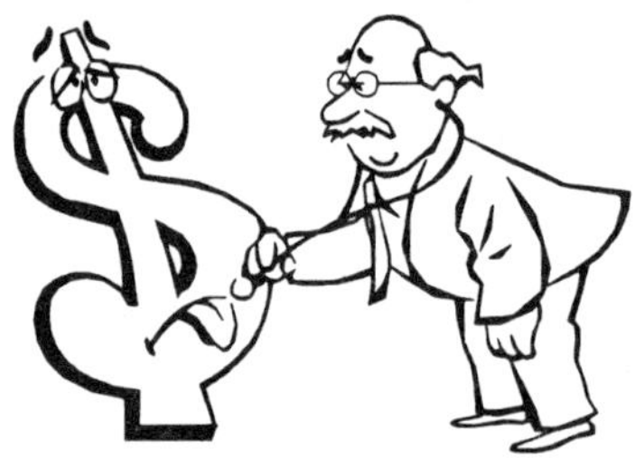

Decide how long you want your mementos to last; learn about paper and how to treat it. You should always purchase the best quality you can afford and print important documents and books on acid-free or archival paper.

You should understand the differences in plastic and page protectors, learn how to de-acidify documents, encapsulate all precious documents with mylar or polyester. You should check your collection periodically to be sure it is in good condition.

In this age of disposable everything, we must be careful not to dispose of our past. It seems that as our world speeds up, more artifacts are left behind. Now is the time for us to come to the aid of our own heritage. It is no small task to find, organize, restore, and preserve our collections, but if this is not done, a valuable legacy will be lost.

GLOSSARY OF TERMS

(The first listed definition is from Webster's Dictionary, Where there is only one explanation or a second one, it is mine according to information in this book.)

Acetate: "Cellulose acetate, a salt or ester of acetic acid. Old page protectors and many packaging materials are made from it." It is a living cotton fiber pulp and acetone formulization known as acetate clear film. Some acetate contains acid and should not be used. All acetate tends to be brittle and cracks easily.

Acetone: "a dangerous, fragrant, volatile solvent." Sometimes used to remove adhesives. Must be used in well vented area.

Acid: The #1 cause of damage to all paper documents and most collectibles. Acids weaken cellulose in paper, board, and cloth, leading to embrittlement and discoloration of these items. Acids are either introduced by internal factors or external factors. Internal factors include the way these items were manufactured and with what. External factors include those things that come in contact with paper, board and cloth items and contaminate them.

Actinic ink: A special acid-free ink that can be used with a dip pen or stamp pad.

Adhesive: "an adhesive substance (glue or cement)."

Alkaline: ". . . of, relating to, or having the properties of an alkali, having a pH of more than 7." Contains no acid.

Alkaline: ". . . of, relating to, or having the properties of an alkali, having a pH of more than 7." Contains no acid.

Alpha cellulose: A form of cellulose derived from cotton. The use of this in the manufacture of paper or board insures stability and longevity.

Antique: "A work of art, piece of furniture, or decorative object made at an earlier period and according to various to Customs [Department and IRS] laws, at least 100 years old." Something that is at least one-hundred years old.

Archival: a term that suggests a material or item is permanent, durable and chemically stable and is therefore safe for the preservation of our keepsakes.

Benzine: "a petroleum based solvent. Sometimes used to remove adhesives." It is oily and dangerous to paperwork.

Blotting paper: a highly absorbent, porous, heavy weight paper that is used for blotting ink and other liquids.

Buffer: "a substance capable in solution of neutralizing both acids and bases and thereby maintaining the original acidity or basicity of the solution." Calcium carbonate is added to a product and prevents any future acid from forming.

Cellophane: "regenerated cellulose in thin transparent sheets used especially for packaging." Highly acidic, will yellow and becomes brittle easily. Old page protectors, usually with black insert sheets, were made from this.

Cellulose: "a polysaccharide of glucose units that constitutes the chief

Cellulose Nitrate: "any of several esters of nitric acid formed by the action of nitric acid on cellulose and used for making plastics." Old film was made from this product and was highly unstable and flammable.

CFC - ChloroFluoroCarbon: "any of a group of compounds that contain carbon, chlorine, fluorine, and sometimes hydrogen and are used as refrigerants, cleaning solvents, and aerosol propellants." Are thought to be dangerous to the ozone.

Cockle: "to wrinkle or pucker." Paper does this if not weighted properly during the drying process.

Collectibles: "something that you collect either for personal enjoyment or as an investment." Any collection that is important to you.

Conservator: "a person that has been given the responsibility of preserving a major collection, as a museum, etc."

Deacidification: "to remove acid, as in neutralization." Treating paper or items containing acid to remove it. This process usually also includes the alkalization of the item by depositing a buffer.

Degradation: "the act or process of degrading, declining to a low state." When something starts to deteriorate it will never be the same.

Desiccant: "a drying agent, as calcium chloride." Usually a small holder which contain a chemical that controls humidity by removing moisture.

holder which contain a chemical that controls humidity by removing moisture.

Durability: the degree to which paper retains its original strength under heavy use.

Encapsulation: "to enclose in or as if in a capsule." To place a document in between two sheets of transparent Mylar D and sealing with double sided tape. It can be easily removed leaving the document in its original condition.

Ephemera: "daily, lasting a day or so, something that dies in a short time, lasting only a few days (like plants, flowers)." To collectors ephemera refers to items such as trading cards, valentines, patterns, paper dolls, post cards, stereo views, etc. Usually items of varying sizes and three-dimensional decorations, or moving parts. Polyethylene bags are often used to protect them.

Fiber content: The types and percentages of fibers used in the manufacture of paper, board and cloth. The quality of the fiber affects the durability and stability of the material.

Fluorescent light: "a tubular bulb having a coating of fluorescent material on its inner surface and containing mercury vapor whose bombardment by electrons provides UV light which causes the material to emit visible light." Less dangerous than daylight or sunlight.

Glassine: "a thin dense transparent or semitransparent paper highly resistant to the passage of air and grease." Highly acidic, will yellow and becomes brittle easily. It contains plasticizer and glycerine. Old negative holders and envelopes were made from this.

Glue: "a hard protein chiefly gelatinous substance that absorbs water to form a viscous solution with strong adhesive properties."

Halogen: "an intense light produced by a gaseous substance called halite." Used in newer household fixtures and lamps. Less damaging then fluorescent light in equal brightness.

Heirloom: "something of special value handed on from one generation to another."

Hide glue: "glue made from animal hides is the traditional glue used in bookbinding and is heated to be used."

Incandescent light: "intense light produced by a heating element made of tungsten." Less dangerous than daylight or sunlight.

Inert: "lacking the power to move, deficient in active properties." Inert materials are ideal for storing most keepsakes.

Keepsake: "something kept or given to be kept as a memento." Any item having family significance, including those having sentimental value to you.

Kraft paper: crude acidic paper, unsafe for long term storage.

Lamination: the process of reinforcing paper and applying a thin plastic protective covering. Most forms of lamination can be dangerous to important items because of the adhesive, the heat used to apply it, and the fact lamination is usually not reversible.

Lignin: "an amorphous polymeric substance related to cellulose that together with cellulose forms the woody cell walls of plants." It is highly acidic. The dangerous part of wood.

Melenex: a brand name of I.C.I. American, Inc. for polyester film.

Memento: "something that serves to warn or remind. Any item that brings back memories."

Methyl Cellulose: a glue that contains no acid. It is in powder form and is mixed with water to use. Completely reversible.

Moisture barrier: a piece of paper, such as waxed paper or polyester film, that prevents moisture from making adjacent paper wrinkle.

Mylar: a brand name for polyester film made by the duPont Company. Mylar D is a product especially designed for use with paperwork and encapsulation. It is very strong and very sharp. It is impervious to heat and cold. Not for use with anything susceptible to static electricity.

Mohawk Superfine: an extremely high quality archival paper available from Gary Shumway (714) 693-8703 or Mohawk Mills.

Non-aqueous: contains no water.

Opaline: "a soft silica powder used as a dry-cleaning agent for paper."

Oxalic acid: "it is caustic and dangerous. It is used to remove unwanted ink from paper or pages of a book."

Papermaking: "the art and science of making paper from raw materials."

Paste: "a mixture used as an adhesive."

pH: "a measure of acidity and alkalinity of a solution that is a number on a scale on which a value of 7 represents neutrality and lower numbers indicate increasing acidity and higher numbers increasing alkalinity and on which each unit of change represents a tenfold change in acidity or alkalinity and that is the negative logarithm of the effective hydrogen-ion concentration or hydrogen-ion activity in gram equivalents per liter of the solution."

Plasticizers: "Chemicals added to plastic products to impart softness, flexibility or rigidity, workability, or stretchability." Sheet plastics containing plasticizers will lift-off type from the page destroying the surface of the document or page. The type transfers to the plastic ruining it as well.

Polyester film: "any of a group of polymers that consist basically of units of ester and are used in making fibers and plastics." Polyester is a common name for the plastic polyethylene terephthalate. Mylar is made from polyester. It is an inert, chemically stable, space age product. It is the most expensive clear material used for page protectors; usually double the cost of other protector materials. It is the only product that will last hundreds of years instead of decades.

Polyethylene: "a polymer of ethylene, a thermoplastic that is resistant to chemicals and moisture, has good insulating properties, and is used in packaging." Household sandwich bags and food storage bags are often made of clear polyethylene. Good for temporary storage of negatives and film.

Polypropylene: "any of various thermoplastic plastics or fibers that are polymers of propylene." Less expensive page protectors are made from this and will not lift type off a page or document. Do not use it in direct contact with film.

Polyurethane: "any of various polymers used in flexible or rigid foams, and resins." Used in paint and finishes.

PVA: Poly-Vinyl Acetate, "a polymerized vinyl compound or plastic." It is an acid-free safe glue, but must be protected from freezing while in its liquid state.

PVC: Poly-Vinyl-Chloride commonly called vinyl, it has a distinctive odor and easily breaks down chemically. It becomes sticky and greasy to the touch. Extremely dangerous to paperwork. Notebooks and page protectors are made from this. Will lift off type from a printed page.

Pressboard: a strong, dense, highly finished paperboard, used for strength and stiffness in thin pieces. Also a smooth board used to dry pages of a book.

Resin: Any of many chemical ingredients used in the making of various types of plastics. The base ingredient in vinyl materials is the polymerized resin vinyl chloride or PVC.

Ream of paper: five hundred sheets of paper.

Relative humidity: "the ratio of the amount of water vapor in the air to the amount which would be present at the same temperature were the atmosphere to be fully saturated."

Reversibility: Being able to undo a process or treatment without affecting or changing the item.

Ring binder: "a binder that holds pages together by means of metal rings that can be opened or closed to add or remove pages." Rings are usually circular (O-rings) but can be D shaped (D-rings).

Scotchpar and Scotchpak: brand names for polyester made by the 3M Company.

Selenium: "a non-metallic element that resembles sulfur and tellurium." It is used in toning a photo to protect prints from sulphur gases.

Sepia: "a print or photo of a brown color resembling sepia, a brownish gray to dark olive brown." Photo prints with a yellowish brown tint that looks old or antique

Silica gel: "colloidal silica resembling coarse white sand and is extremely absorbent." Will keep your collection dry. Loose silica gel should not be placed in direct contact with photos or negatives, it will scratch them badly. It should not touch any keepsake directly.

Sizing: "any of various glutinous materials made of glue, flour, varnish, or resins and used for filling the pores in surfaces of paper, etc." Chemicals added to paper that make it less absorbent so that inks will not bleed.

Slipcase: a container specifically made for a book that fits perfectly around it, and the book slips in and out very easily.

Thymol: "a crystalline phenol of aromatic odor and antiseptic properties found in thyme oil or made synthetically and used chiefly as a fungicide and preservative." Toxic.

Tissue: very light paper used for interleaving and mending.

Toluene: "a solvent made from petroleum products, similar to benzene but less volatile and toxic." Used to remove adhesives.

Transmigration: refers to the resulting damage potential caused by the transfer of chemicals by contact, from either the acidity of paper or plasticizer content of page protectors. Transmigration potential increases with heat. Also called acid transfer.

Triacetate: "a textile fiber consisting of cellulose that is completely or almost completely acetylated." Older holders for film and prints were made from triacetate.

Tyvek: a special product made by duPont Co. from olefin. It has some advantages of both paper and plastic. It is acid-free and very strong. Boards and tapes are made from it.

Ultraviolet radiation (UV): "electromagnetic radiation of wavelengths shorter than visible light." It can cause fading and other damage to books, paper and cloth if they are exposed to UV for periods of time.

Vinyl: a word commonly used to describe a number of different plastics some of which are not safe for preservation.

Window: "an area cut out of a piece of mat board, slightly smaller or larger than a document or photo being displayed."

LIST OF SELECTED SUPPLIERS

Specific products mentioned in the text are given not because they are necessarily the best of their kind but because they have been found to be satisfactory by me or recommended by others. They could be a starting point for you in your search for the particular items most suitable to your needs.

(Page protectors & photo pages)
20th Century Plastics
205 South Puente Street
Brea, CA 92621
(800) 767-0777
(call for free sample & catalog)

Conservation Materials Ltd.
1165 Marietta Way,
P.O. Box 2884
Sparks, NV 89431

Conservation Resources
International, Inc.
1111 North Royal St.
Alexandria, VA 22314
(703) 549-6610

(Bookbinding supplies)
Gane Brothers & Lane, Inc.
4697 E. 48th St.
Vernon, CA 90058
(213) 582-8331

Gaylord Brothers, Inc.
P. O. Box 4901
Syracuse, NY 13221
(315) 457-5070

The Hollinger Corp.
3810 S. Four Mile Run Dr.
Arlington, VA 22206
(703) 671-6600

I.C.I. America
Plastics Division
Wilmington, DE 19897
(302) 575-3305

Light Impressions
439 Monroe Ave.
P. O. Box 940
Rochester, NY 14603-0940
(800) 828-6216
(Beautiful catalog available!)

Preservation Emporium
P.O. Box 226309, Dept. 93
Dallas, TX 75222-6309
(800) 442-2038

Restoration Source
P. O. Box 9384
Salt Lake City, UT 84109-0384
(801) 278-7880

Transilwrap Co.
135 English St.
Hackensack, NJ 07601

University Products, Inc.
517 Main Street
P. O. Box 101
Holyoke, MA 01041-0101
(800) 762-1165
(beautiful catalog available!)

Wei T'o Associates, Inc.
Drawer 40
Matteson, IL 60443
or
P. O. Box 352
224 Early St.
Park Forest, IL 60466
(312) 748-2995

LIBRARY SUPPLY COMPANY

Brodart
P. O. 3037
1609 Memorial Avenue
Williamsport, PA 17705
(800) 233-8959

BIBLIOGRAPHY

Eric Burdett, *The Craft of Bookbinding, A Practical Handbook* (London: David & Charles, 1975).

Anne F. Clapp, *Curatorial Care of Works of Art on Paper: Basic Procedures for Paper Preservation.*

Douglas Cockerell, *Bookbinding and the Care of Books* (New York: Appleton, 1902). Reprinted many times.

Conservation of Photographs. Kodak

Jane Greenfield, *Books, Their Care and Repair.* (Yale University Library; 1983)

William G. Hartley, *When in Doubt, Don't Throw it Out.* Ensign, April 1986, p. 52-55.

Dard Hunter, *Papermaking. The History and Technique of an Ancient Craft.* (New York: Knopf, 1947), 2nd ed.

Randall K. Mehew, *Organizing Families & Reunions.* (Keepsake Paperbacks, 1994), 125 p.

Ilene Chandler Miller, *Publishing a Family History from P.A.F. A Five Step Guide.* (Shumway Family History Services, 1994), 103 p.

Newsprint and Its Preservation. *(Washington, D.C.: Library of Congress, Preservation Office).*

Northeast Document Conservation Center, *Storage Enclosures for Photographic Materials.* (Andover, MA: Technical Leaflet #001-0008).

Polyester Film Encapsulation, produced by the Library of Congress Preservation Office, Washington D.C. (Available from the Superintendent of Documents, U. S. Govt. Printing Office, Washington, D.C., 20402, Stock #030-000-00114-1.

Wallace and Bousfield, *Successful Family Organizations, Record-Keeping, and Genealogy in Family Activities..* (LL Company, 1978).

Peter Waters, *Water—Damaged Library Material.* (Washington D.C.: U.S. Govt. Printing Office, Library of Congress; 1979)

Robert A. Weinstein and Terry Book, *Collection, Use, and Care of Historical Photographs.* (Nashville, TN: Amer. Assn. for State & Local History, 1977).

Laura S. Young, *Bookbinding & Conservation By Hand, A Working Guide* (N.Y, & London, R. R. Bowker Company, 1981), 273 p.

There are dozens of books written about paper, books and photographs and should be consulted for more detailed information.

INDEX

Items available from the author:

Preserving Family Keepsakes Do's and Don'ts. (Shumway Family History Services, 1995), soft cover, 178 p., $18.50 plus $3.50 s&h, add tax where applicable.

Publishing a Family History from P.A.F. A Five Step Guide. (Shumway Family History Services, 1994), comb bound, soft cover, 103 p., $15 plus $3.50 s&h, add tax where applicable.

***Title Page Templates** for WordPerfect 5.1, 6.0, 6.1, WPWin, & WordWin:* Beautiful ready-made title pages for documents and family histories created in WordPerfect or Word for Windows. They are templates, all you need to do is pull them up into your word processor as you would a file, change the names and dates, and you have instant, professional-looking title pages! Available for IBM & MAC. $10.ea. +tax+ $1 s&h.

- *Special Collection:* includes 10 title pages as shown in book above, + 2 bonus pgs.
- *Collection #1:* includes 10 finished titles pages plus 3 bonus pages. These are less ornate borders and boxes.
- *Collection #2:* includes 10 finished titles pages plus 2 bonus pages. These are ornate/fancy borders & boxes.
- *Collection #3:* includes 10 finished titles pages. These are popular assorted designs.
- *Trees Collection:* A collection of decorative title pages which include TREES, especially designed for use with your family tree.
- *Etc. Collection:* Contains 9 oval borders plus 9 styles of "My Family Tree" to mix and match as you want. Included are 2 templates "Wanted" and Reward" to use as correction copy title pages. Also includes graphic text breaks and headings.
- *L.D.S. Collection:* 10 finished title pages which include designs of interest to LDS members.

***Clip Art:** "The Child":* 29 delightful 19th century children in PCX, TIF & PCT format.

ORDER FORM

Name:____________________________________

Address:__________________________________

City:_______________State_____Zip________

Telephone: ( ) ___________

IBM____ MAC____Word Processor___________

Version:_____

Keepsake Publishing
9712 Mirage Circle
Garden Grove, CA 92644
(714) 636-3536

(Send check or money order, include s & h and tax)
Add $1. extra for Canada